The World Bank and Urban Development

As one of the world's most powerful supranational institutions, the World Bank has played an important role in development ideology and practice since 1946. Although the World Bank has been involved in urban lending for more than three decades, *The World Bank and Urban Development* is the first book-length history and analysis of the Bank's urban programs and their complex relationship to urban policy formulation in the developing world. Through extensive primary research, including interviews with World Bank and government officials, and through an exploration of factors internal and external to the Bank that have impacted its urban development agenda, this groundbreaking work addresses four major gaps in the literature:

- the political and economic forces that propelled the reluctant World Bank to finally embrace urban programs in the 1970s;
- how the World Bank fashioned its general ideology of development into specific urban lending projects and how those programs, in turn, eventually translated into urban policy in developing countries;
- trends and transitions within the Bank's urban agenda from its inception to the present;
- the World Bank's historic and contemporary role in the complex interaction between global, national, and local forces that shape the urban agendas of developing countries.

The book also examines how protests from NGOs and civic movements, in the context of globalization and neo-liberalism, have influenced World Bank policies from the 1990s to the present. The institution's attempts to restructure and legitimate itself, in light of shifting geo-political and intellectual contexts, are considered throughout the book.

Edward Ramsamy is Assistant Professor of Africana Studies and a member of the graduate faculties of geography and urban planning/policy development at Rutgers University.

Routledge Studies in Development and Society

The World Bank and Urban Development

From projects to policy

Edward Ramsamy

Routledge
Taylor & Francis Group

LONDON AND NEW YORK

First published 2006
by Routledge
2 Park Square, Milton Park, Abingdon, Oxfordshire OX14 4RN

Simultaneously published in the USA and Canada
by Routledge
711 Third Avenue, New York, NY 10017

Transferred to Digital Printing 2007

Routledge is an imprint of the Taylor and Francis Group, an informa business

First issued in paperback 2015

© 2006 Edward Ramsamy

Typeset in Gill Sans and Baskerville
by Prepress Projects Ltd, Perth, UK

British Library Cataloguing in Publication Data
A catalogue record for this book is available from the British Library

Library of Congress Cataloging in Publication Data
A catalog record for this book has been requested

ISBN 978-0-415-34439-5 (hbk)
ISBN 978-1-138-98735-7 (pbk)
ISBN 978-0-203-49408-0 (ebk)

This book is dedicated to all my family, but especially to my parents, Puripoornam and John Ramsamy, who created spaces of meaning and hope

Contents

Figures

Tables

Acknowledgments

In the completion of this book, I have benefited greatly from the advice, assistance, and kindness of many people. The intellectual inspiration for this project came from a variety of sources. I completed my undergraduate education at the University of Durban-Westville (UDW) in apartheid South Africa during the politically volatile 1980s. There, Brij Maharaj and Dhiru Soni introduced me to critical and progressive geographic thought against a deep current of cynicism, skepticism, and hostility from establishment social scientists at UDW. Their courage in that environment allowed me to follow my intellectual curiosity to a variety of new paradigms, some of which eventually found their way into the present work. Soni especially encouraged me to pursue graduate studies in the United States. Without his impetus, this book might never have been. Maharaj always posed challenging questions, ever since my undergraduate days. I am truly privileged to have had extended conversations with him over the years. His provocative ideas and principled politics have inspired me both intellectually and personally. I was also fortunate to have worked as a research assistant for Vishnu Padayachee, one of South Africa's leading economic historians. Padayachee's keen mind and meticulous approach left a permanent imprint on me as an aspiring researcher. The generosity of these and numerous other figures at UDW has been inestimably important to my intellectual development.

It was an honor for me to write my doctoral dissertation under the supervision of Susan Fainstein at Rutgers University. She combined thoughtful criticism with encouragement and support, without which I might have lost focus along the way. Her continued guidance is of great value to me. Neil Smith's monumental contribution to geographical thought greatly shaped my own evolving conceptualizations of space and society. I owe much to his insightful reading of this book, especially his comments on the World Bank and social movements. Cornel West and D. Michael Shafer gave graciously of their time to discuss with me some trends in cultural studies and development theory, respectively, when the idea of this work was still germinating.

Colleagues in Africana Studies, Geography, and the Edward J. Bloustein School of Public Policy at Rutgers University extended themselves in many ways during the gestation of this book. Gayle T. Tate has been a true comrade. Her exuberant personality, incomparable wit, knowledge of politics, and sense of shared struggle have always been a source of motivation, especially when things did not seem to go as planned. I am grateful to Walton Johnson, who has shown keen interest for this project from its inception. His expertise on southern African issues has helped me to formulate my own ideas. Johnson, Bria Holcomb, and Don Kruekeberg read the entire manuscript and posed brilliant questions. Holcomb teased out points that required elaboration while Kruekeberg applied his methodical reasoning to the arguments. Without their constant nudging, this book might not have been finished. I sincerely appreciate Leonard Bethel's and Kim Butler's encouragement of my scholarly endeavors as well as their understanding of the burdens of teaching. Secretaries Barbara Mitchell and Janice Gray deserve thanks for their administrative assistance. I also benefited from interactions with other individuals at Rutgers and the larger academic community, including Hooshang Amirahmadi, John Butler Adam, James Blaut (late), Stephen Bronner, Ethel Brooks, David Burns, Pedro Caban, Joan Carbone, John Cooney (late), Gerald Davis (late), Salah El-Shakhs, Prosper Godonoo, Allen and Peggy Howard, Cora Kaplan, Cindy Katz, Margaret Klawunn, Anthony Lemon, George Levine, Greg Metz, Bruce Robbins, Eve Sachs, Biswapriya Sanyal, Rick Shain, Himanshu Shukla, Mala Singh, Prem Singh (late), Meredith Turshen, and Ben Wisner, among many others. I am grateful to all of them.

This book was made possible by several awards, fellowships, and grants. Fellowships from the Center for the Critical Analysis of Contemporary Culture (CCACC) and the Social Science Research Council (SSRC) enabled me to do field work in Zimbabwe. Deans James Reed and Marie Logue provided additional assistance that permitted me to attend to important conferences that were invaluable in fine-tuning the book. The Dean's Office of the Faculty of Arts of Sciences at Rutgers University also provided financial support that brought this project to fruition. I am particularly grateful to Dean Douglas Blair for his effort to insure that I would have the space to pursue my ideas.

Officials from the World Bank, United States Agency for International Development (USAID), Harvard Institute for International Development, and the Zimbabwean government agreed to be interviewed by me and took time from their demanding schedules to explain the programs of their institutions. I am particularly grateful to Kenneth Bohr, Michael Cohen, Douglas Keare, and Edward Jaycox, pioneers in the World Bank's urban programs. Discussions with Colleen Butcher, Chris Mafico, Beacon Mbiba, and Diana Patel enhanced my understanding of local politics in Zimbabwe. Although these individuals may not necessarily agree with my conclusions, they, nevertheless, happily entertained my curiosity. I also consulted numerous libraries at the World Bank (Washington and Zimbabwe), USAID, the UN, the University of Zimbabwe, the Ford Foundation, and the Library of Congress. I owe thanks to the staff of these institutions for their expert assistance. Mary Fetzer, the former government resource librarian at Rutgers University, was especially helpful in navigating the vast and complex world of government documents, of which I made extensive use in this project. I am also grateful to Mary Marshall Clark, the Associate Director of the Oral History Research Office at Columbia University, for guiding me through the oral history collection on the World Bank. I am indebted to Emmanuel Hove, whose hospitality in Zimbabwe made it possible for me to meet and interview many officials and members of the public in that country. Extended conversations with him further enabled me to develop a coherent understanding of what I had observed and learned.

I would like to thank Andrew Mould, senior editor at Routledge, for believing in this project and offering valuable suggestions on directions it might take. Zoe Kruze, Emma Hart, Andrew R. Davidson, and Helen MacDonald deserve thanks for their contributions toward the book's production.

The social and personal support offered by friends along the way was integral to the writing of this book. My dear friend Olubayi Olubayi went out of his way to help, academically and personally, at critical moments during several phases of this work. His eclectic intellectual interests always made for great conversation while his understanding of the academic life and the research process prodded me along during periods of great uncertainty. Prem Govender and Stanley Arumugam provided comic relief and rational antidotes to prevailing dogmas and humbugs. Emeka Akaezuwa somehow found time to help even when there was no time. Amin Khadr's musings on aesthetics never failed to amuse or prompt further musings. Dushen Chetty, Anita Lalloo, Denniston Bonadie, Wendel Thomas, and Joan Leung Lo-Hing have always been good company. Niraj, Reena, Akshay, and Asheeta Bothra often provided me with refreshing respite from my work when I was weary. I appreciate their friendship very much.

I owe my academic career to my immediate and extended family. My parents, Puripoornam and John Ramsamy, to whom this book is dedicated, and my siblings, Shanthi, Silvia, and Emmanuel, each sacrificed a great deal to enable me to pursue tertiary education. Without their unfailing love, prayers, sacrifice, and material support, I could not have written this book. Herby and Devarakshanam Govinden and Jonathan Jack always took a keen interest in my educational development and sparked my love of learning at an early age. Uncle Herby's achievements as the first Indian to gain a PhD in science in South Africa inspire me daily, as does his ongoing spirit of intellectual inquiry, exemplified by his current pursuit of a second doctorate in cosmology. Pavadaisamy and Kalpakam Ramachandran, my parents-in-law, opened up their home to me and continue to tolerate my clutter. The gracious hospitality of the Tamilarasans in the UK during research trips there will not be forgotten. Anitha Ramachandran, officially my sister-in-law, adopted me as her own brother. With an undergraduate degree in biology and graduate training in literary theory, she offered unique and refreshing insights into a public policy manuscript. Her moral support helped me to see this project through.

I owe an enormous debt to my wife Kavitha, whose love, friendship, and passion for ideas sustain me in ways too numerous to mention here. This book is the result of her persistent faith in my endeavors. Even as she juggled her own teaching, research, and motherhood, she read the entire manuscript and offered incisive criticisms that enriched the final product. My son, Selvan, has enabled me to see the world with new eyes. I truly have been blessed by their presence in my life.

Abbreviations

AEA	American Economic Association
CABS	Central African Building Society
CDC	Commonwealth Development Corporation
CDF	Comprehensive Development Framework
CDLF	Central Development Loan Fund
CHUDSA	Conference on Housing and Urban Development in Sub-Saharan Africa
DED	Development Economics Department
ESAP	Economic Structural Adjustment Program
FAO	Food and Agriculture Organization
FBS	Founders Building Society
GATT	General Agreement on Tariffs and Trade
GDF	General Development Fund
GLDF	General Loan Development Fund
HGF	Housing Guarantee Fund
HIGP	Housing Investment Guarantee Program
IBRD	International Bank for Reconstruction and Development
IDA	International Development Association
IDB	International Development Bank
ILO	International Labor Organization
IMF	International Monetary Fund
IRD	integrated rural development
LDC	less developed country
MCNH	Ministry of Construction and National Housing
MDB	multilateral development bank
MDC	Movement for Democratic Change
MFEPD	Ministry of Finance, Economic Planning and Development
MLGRUD	Ministry of Local Government, Rural and Urban Development
MPCNH	Ministry of Public Construction and National Housing
MLGTP	Ministry of Local Government and Town Planning
NATO	North Atlantic Treaty Organization
NGO	non-governmental organization

NHF	National Housing Fund
NIC	newly industrializing country
OAU	Organization of African Unity
PUPS	permanent paid-up shares
SAL	structural adjustment loans
SECALs	sectoral adjustment loans
SLC	Staff Loan Committee
SUNFED	Special United Nations Fund for Economic Development
TNDP	Transitional National Development Plan
TOD	Technical Operations Department
UDI	Unilateral Declaration of Independence
UNCHS	United Nations Centre for Human Settlements
UNDP	United Nations Development Programme
UNEDA	United Nations Economic Development Administration
UNESCO	United Nations Educational, Scientific and Cultural Organization
UNISD	United Nations Institute for Social Development
USAID	United States Agency for International Development
WHO	World Health Organization
WSF	World Social Forum
ZAAT	Zimbabwe Association of Accounting Technicians
ZANU	Zimbabwe African National Union
ZANU-PF	Zimbabwe African Union – Patriotic Front
ZAPU	Zimbabwean African People's Union
ZESA	Zimbabwe Electricity Supply Authority
ZIMCORD	Zimbabwe Conference on Reconstruction and Development

Introduction

As one of the most powerful multilateral development institutions in the world, the World Bank needs no introduction. It is a complex organization with the mandate to enable the economic development of Third World countries. Staffed by 800 economists, more than 3,000 engineers, technical experts, and other professionals, and allotted a budget of US$25 million for research alone, the World Bank overshadows nearly all other entities working on development. An annual loan budget of about US$15–20 billion enables the Bank to translate ideology into policy and exercise considerable influence in the affairs of developing countries. This observation has led commentators to conclude that "the Bank has more to say about state policy than many states" themselves (George and Sabelli 1994: 1).

The end of World War II brought a host of supranational organizations, including the World Bank and International Monetary Fund (IMF), into the global constellation of governance and finance. Since then, these organizations have had a say in virtually every aspect of domestic policy in less developed countries (LDCs), from development, to trade, to the environment. Today, in the context of deepening globalization, the World Bank appears as a mighty leviathan that eclipses the state in developing countries. Indeed, many weak states are vulnerable to the Bank's will, but even the stronger states must also contend with the Bank's hegemonic leadership.

This book focuses on the Bank's role in an aspect of domestic policy in developing countries: urban development. Urbanization was a powerful socio-spatial force throughout most of the world during the twentieth century. Compared with the West, high and sustained rates of urbanization in developing countries have produced enormous problems for infrastructure and service provision (Berry 1981; Mabogunje 1981; Gilbert and Gugler 1982; Linn 1983). In their attempts to deal with burgeoning urbanization, Third World national governments tried to discourage the phenomenon by mitigating the push factors for migration to cities (El-Shakhs 1972; Brennan and Richardson 1986), altering the location of investment, and decentralizing economic activities to smaller urban areas (Lipton 1976; Forbes and Thrift 1987). However, such policies proved to be futile and ineffective after wasting already scarce resources (Richardson 1987a; N. Harris 1989). Given the inevitability of hyper-urbanization under uneven capitalist development, policy-makers have had to confront its consequences, especially the alarming growth in the number of urban poor and their lack of access to basic services and infrastructure. A major consequence of rapid urbanization in the developing world is the prolific growth of slums and squatter settlements. This informal housing usually lacks, or has limited access to, clean water, sewerage systems, and electricity. In some instances, these dwellings constitute the housing of over 60 percent of the total urban population (Potter 1985; United Nations 1996; World Bank 2000a).

However, neither the exceptional growth of Third World cities nor the deteriorating living standards of their urban poor was a major concern for international development agencies until the early 1970s. As poverty was regarded as an overwhelmingly rural problem, efforts were largely directed toward addressing rural underdevelopment. It was also believed that investment in the provision of social services and infrastructure for Third World cities would simply reinforce an existing "urban bias" that was already depriving rural areas of much needed capital and human resources (Stren 1994: 4). Thus, during the 1950s, for example, the World Bank's lending focused solely on public utility projects, such as electric power and transportation. The Bank did not allocate funds for socially oriented development programs, such as health, education, and housing, for fear of promoting welfarism. In the late 1960s, however, a variety of internal and external forces influenced the institution to consider the poverty of developing countries. As part of this evolution in its thinking, the World Bank began to pay closer attention to urban issues.

Mounting criticism resulting from increasingly visible physical manifestations of poverty by the mid-1960s (Abrams 1964; Ward 1965) eventually prodded the Bank to confront the problem. The World Bank reluctantly considered approaches to alleviating urban poverty and revised its antagonistic policy toward funding urban projects in 1972.

Most of the Bank's early interventions in poverty alleviation aimed to extend basic infrastructure to squatter settlements. After initiating an urban lending program in 1972, the Bank used its ample financial resources to assist with housing provision in the Third World. Sites-and-services and squatter upgrading were promoted by the Bank for a decade as acceptable solutions to the Third World's housing problems. However, during the 1980s and 1990s, the Bank shifted its emphasis from project-based lending to macroeconomic management and "capacity-building," concentrating on increasing productivity. As a consequence, its urban focus moved away from investing in specific projects, such as physical dwellings, toward the reform of urban finance and eventually toward privatization (World Bank 1991a, 1993a, 2000a).

Although the World Bank has been involved in urban lending for more than three decades, there is a paucity of analysis on the history and evolution of the Bank's approach to urban development. Most of the literature on the Bank's urban programs has been produced by the Bank itself. While agricultural lending, structural adjustment, and the institutional structure of the World Bank have been the subjects of many studies (Torrie 1983; Havnevik 1987; Lipton and Paarlberg 1990; Miller-Adams 1999), similar analysis of the World Bank's urban programs is lacking. For example, the Brookings Institution's two-volume study (Kapur *et al.* 1997) on the first fifty years of World Bank programs dedicates only about thirty of its 2,000 pages to the Bank's urban initiatives. While a few studies examining urban programs in specific cities do exist (Domicelj 1988; Pugh 1988, 1989a; Campbell 1990; Guarda 1990), no major works have been produced on the history of the World Bank's urban program, the factors that led to the establishment of the Bank's urban division, the relationships between the Bank's urban programs and its broader development objectives, the impact of the Bank's urban lending philosophy on urban development policy in general, or trends in the Bank's urban lending program.

This book addresses these gaps in the literature through an examination of factors internal and external to the Bank that have influenced its urban development agenda. The book describes how the World Bank became involved in urban lending, how it fashioned ideas into projects and programs, and how it eventually translated programs into specific policies. The trends and transitions in the Bank's urban lending programs are then discussed. Undoubtedly, given the size of the institution, there is a plurality of views among program directors, management, and general staff regarding how to approach the problem of urban development. Nevertheless, it is possible to identify a convergent set of views that characterizes its thinking on urban development at different historic moments. Finally, the book also explores how the Bank's urban programs have intersected with the national development agenda of a specific developing country, Zimbabwe.

One of the main arguments of the book is that the establishment of the World Bank's Urban Division in 1972 and the subsequent shifts in its urban agenda were not purely outcomes of bureaucratic decisions within the Bank based on technical evaluations of projects; instead, I argue that they are responses to geo-political and intellectual trends within and outside the Bank. Next, using Zimbabwe as an example, I demonstrate that, although shifts in the World Bank's urban agenda have important consequences for domestic policies of developing countries, the Bank is more than an agent of domination; it is, in fact, a hegemonic, supranational actor that articulates with national states and other actors to define the course of development in LDCs. Furthermore, I contend that dualistic approaches that view the Bank's relationship with developing countries as either positive or negative are inadequate for understanding this powerful institution's role in an increasingly complex and global world.

The observations and arguments of the book are organized into seven chapters. In Chapter 1, "Theorizing the World Bank and development," I overview major theories of development and relate them to the Bank. I identify some weaknesses that render these theories inadequate for conceptualizing the Bank's role in policy formulation for the Third World. I then present a framework that avoids the reduction of the World Bank's hegemony to domination, on one hand, and the overdetermination of the "local" by the "global," on the other hand, by paying particular attention to the politics of scale and the role of national states. Chapter 2, "Toward social lending: shifts in the World Bank's development thinking," overviews the Bank's early loan programs and identifies factors that propelled the Bank to recognize social concerns in developing countries. The Bank's eventual embrace of urban development is shown to be part of an overall shift toward the acceptance of poverty alleviation as a basis for lending. These shifts are discussed in relation to Robert McNamara's presidency (1968–81), which deeply influenced the Bank's thought and action during this period. In Chapter 3, "The search for an urban agenda at the World Bank," I present an overview of the Bank's early urban initiatives and examine the intellectual and policy trends that led the Bank to alter its once ambivalent attitude toward urban lending. Chapter 4, "The fall of poverty alleviation: the politics of urban lending at the World Bank," is about the Bank's transition from poverty alleviation toward a more fiscally conservative stance, and the implications of this shift for the Bank's urban agenda. The bureaucratization of the Bank's urban concerns is traced and situated in relation to the new political climate of the 1980s, which emphasized structural adjustment and, later, governance. Based on an examination of the Bank's shift in housing policy during these years, I show how the urban agenda of this period was pressured to keep pace with emerging political trends, both internal and external to the Bank. Chapter 5, "Beyond global and local: a critical analysis of the World Bank and urban development in Zimbabwe," illustrates the complex interaction between the World Bank and national actors, using the Zimbabwean experience. This chapter discusses the elements of the Bank's key urban initiatives in Zimbabwe and assesses whether the Bank's urban programs were able to meet the needs of the urban poor, as claimed by the Bank. Chapter 6, "Globalization, neo-liberalism, and the politics of the World Bank's current urban agenda," and the "Conclusion" (Chapter 7) examine how globalization and neo-liberalism have reinforced conservative World Bank policies from the 1990s to the present. Protest from a variety of civic groups and non-governmental organizations has pressured the Bank to revise its hard-line stance of the 1980s. The Bank's attempts to restructure itself in light of these shifting contexts are considered throughout the book.

Chapter 1

Theorizing the World Bank and development

A major controversy associated with the World Bank is whether its programs promote or thwart development in the Third World. While the Bank's charter affirms that it is a purely economic institution whose primary function is to provide loans for specific projects, the Bank wields considerable power as a supranational agency shaping policy in many developing countries. Over the decades since its inception at Bretton Woods, the World Bank has become an increasingly hegemonic player in development. Its capacity for realpolitik enables it to move with great ease through the geo-political complexity of the world. Whether it is promoting broad policy "reform," such as structural adjustment, fiscal management, or governance, or implementing particular projects in specific locations the Bank repeatedly has proven itself to be a "glocal" actor.[1]

The aim of this chapter is to develop a framework that theorizes the process by which the World Bank affects policy choices in developing nations. The chapter consists of three parts. The first part overviews major development theories and relates them to the World Bank. The second part identifies some weaknesses that render these theories inadequate for conceptualizing the Bank's role in the policy-making process in the developing world. In the third part, I present an alternative framework that addresses the shortcomings of development theory and captures the uneven and scaled articulation between the World Bank and LDCs. Existing literature on the Bank's relationship with developing countries tends to view the interaction in binary terms, as either positive or negative. The Bank is seen as either a catalyst for growth for developing countries or an instrument of domination that stifles national development, promotes dependency, and increases vulnerability; in fact, the relationship is more complex. Based on the premise that the world political economy both shapes and is shaped by individual states and modes of regulation, I suggest that a more nuanced reading of the Bank's role in development is needed, one that avoids the reduction of hegemony to domination and the overdetermination of the local by the global.

Theories of development and the World Bank

The World Bank and the modernization paradigm

During the 1950s and 1960s, development thinking and planning were dominated by the modernization approach. With its intellectual roots in the writings of Spencer, Weber, Parsons, and Bentham, modernization theory saw development as a gradual, evolutionary process involving various stages and transforming all societies from traditional to modern. As societies modernized, they were supposed to develop complex economies, institutions, bureaucracies, and divisions of labor that enabled them to meet their production and consumption needs. In keeping with the neo-classical economic tradition, modernization theory advocated integration into the global capitalist system, economic growth, and Western liberalism as a way of achieving development.

Influenced by evolutionary theory, modernization theory posited that social change occurs as societies move linearly from traditional to advanced, implying that the movement represents progress, civilization, and development. Second, the evolution from a simple, primitive society to a complex, modern one is seen as a long, slow, incremental, but irreversible process. Third, the homogenization of societies caused by modernization is said to enable effective economic linkages. Cultural convergence among developing societies in the form of Westernization also homogenizes them, with the assumption that following the European and American examples might lead to levels of economic prosperity enjoyed by the West (Rostow 1964).

Modernization theory also borrows from functionalism. In this regard, modernization is seen by its proponents as a comprehensive program, affecting all social functions and leading to changes in industrialization, urbanization, social organization and differentiation, and participation. Modernization is also assumed to replace traditional values with modern ones. Following Parsons (1951), modernization theory holds that, if they are to survive, societies have to adapt to their environments, attain specific goals, integrate within themselves, and transmit modern values from generation to generation through the economy, government, and institutions.

Applying neo-classical economics to urbanization, Berry (1970) and others promoted the view of "growth impulses" that diffuse down the urban hierarchy, much like the diffusion of ideas. In explaining urban development in the Third World, Friedmann (1966: 35) argued that "where economic growth is sustained over long periods, its incidence works toward the progressive integration of the space economy" such that, eventually, a functionally interdependent system of cities would emerge. Economic development, according to Friedmann, will ultimately lead to the convergence of regional incomes and welfare.

The modernization perspective has greatly influenced how the Bank's relationship with developing countries is theorized and represented. Rostow's (1960) ideas, Rosenstein-Rodan's "big push" model (1957), and the "two-gap" growth model developed by Chenery and Strout (1966) all highlighted the theme of "gap-ism" between the developed and developing worlds. These perspectives advocated a significant increase in productive investment, the development of a manufacturing sector, and the creation of a social, political, and institutional framework that facilitates the mobilization of capital. During the 1950s and 1960s, there was a strong belief that external inducements were crucial in bridging the gap between the developed and developing worlds.

While the World Bank was not as important a player in the international development scene during the early post-war period as it is in the contemporary period, it nevertheless provided assistance for projects that complemented the logic of the modernization school of thought. The evolutionary perspective of the modernization approach greatly influenced the Bank's own thinking on development and led the institution to see itself as a catalyst to economic growth. From 1945 to about 1970, the Bank firmly believed in external capital investment as a "locomotive" to push traditional societies to modernize themselves (Shihata 1991). The overt functionalism of the modernization approach is also apparent in the Bank's world view. Investment in key sectors such as mining and manufacturing was seen as enabling accumulated wealth to "trickle down" and reach the poor, who were eventually to benefit from it in the long run. Urban infrastructure loans were intended to support projects that would promote the "growth impulse," whose benefits would then diffuse into the entire economy.

Following the logic of the modernization paradigm, the Bank advocated that in order for private initiative and investment to occur there needs to be an adequate complement of public overhead capital in the form of railways, roads, power plants, ports, etc. Production will expand, it was assumed, once infrastructure and a satisfactory climate for private investment are in place. The Bank's role, then, was to provide development assistance (both material and technical) to meet these requirements. This view is reflected in the Bank's sixth annual report:

> It is only natural that, except for the early reconstruction loans, the Bank's lending operations have been concentrated in the field of basic utilities. An adequate supply of power, communications and transportation facilities is a pre-condition for the most productive application of private savings in new enterprises. It is also the first step in the gradual industrialization and diversification of the underdeveloped countries. These basic facilities require large initial capital outlays, which, because of the low level of savings and the inadequate development of savings institutions, often cannot be financed wholly by the countries themselves. Moreover, most of the machinery and equipment used in the construction of these facilities

must be imported. Therefore the resources of the Bank are called upon to provide the foreign exchange necessary for the building of these vitally important facilities.

(World Bank 1950–1: 14)

Dependency/world systems approaches and the World Bank

Development theorists, such as Hayter (1971, 1981), Payer (1974, 1982), and Wellings (1982), have applied the logic of dependency theory in examining the relationship between the programs of the World Bank and development. This school of thought argued that, instead of promoting socioeconomic development in Third World countries, interventions by organizations such as the World Bank lead to increased dependency and international vulnerability. The dependency and world systems arguments draw inspiration from traditions that challenged the intellectual dominance of modernization theory and the political ideologies of the capitalist West. In contrast to advocates of modernization, who view the Third World's interaction with the world economy as beneficial, dependency theorists regard it as a constraint. Two major theoretical assumptions characterize the dependency school. First, the international political economy is conceptualized as a hierarchically ordered system of dominance. Second, the development process in the periphery is a function of the way in which it is incorporated into the international division of labor. Since development in the core causes underdevelopment in the periphery, according to this perspective, external forces are primarily responsible for the distortions that characterize the economies of the developing world (Frank 1967; Amin 1972; Rodney 1972).

At any given time, the picture of the world, according to Frank, consists of:

> a whole chain of metropolises and satellites, which runs from the world metropoles down to the hacienda or rural merchants who are satellites of the local commercial metropolitan center but who have peasants as satellites.
>
> (Frank 1967: 146–7)

This formulation formed the basis of Castells' (1977) more systematic explication of dependent urbanization. His main quarrel with the modernization paradigm was that it did not explain how urbanization unfolded under dependent capitalism. While he acknowledged that city growth differed in form and nature in various parts of the Third World, the processes of urban development, nevertheless, ought to be understood as the expression of global capitalist political economy.

The expansion of the global economy into peripheral areas leads to a development dynamic in which a few large cities act as trade centers in the web of colonial and neo-colonial exploitation. This results in a form of urbanization that leads to urban primacy, regional inequality, and centralization of political and economic power within cities (El-Shakhs 1972). Chase-Dunn (1984: 115) explains the role of the dependent city in the world system: "Peripheral primate cities are nodes on a conduit which transmit surplus value to the core and domination to the periphery, while primate cities in the core receive surplus value and transmit domination." Dependent urbanization is, thus, characterized by high levels of unemployment, material inequality, poverty, and technological and financial dependence. Although there are significant variations in the theme of dependency, this school of thought contends that it would be impossible to comprehend most economic, social, and political phenomena in Third World regions unless they are theorized as being structurally connected to the economic and political systems of the advanced capitalist countries.

Wallerstein's (1974, 1979, 1980) world systems perspective complements the dependency argument. It agrees that development problems in the periphery are caused by global capitalism. However, while the dependency approach applies mainly to former European colonies, the world systems approach regards the global political economy as the unit of analysis and avoids the internal versus external dichotomy from the dependent country's viewpoint. Wallerstein argued that states are inappropriate units of analysis for studying economic and political development and called for a wide-angled approach that conceptualized the world political economy as an integrated system. Wallerstein identified three basic elements in the world system:

1   there is a single world market, which is capitalist;
2   the world political structure is mediated through a competitive inter-state system; and
3   while the world economy is constructed by geographically integrating a vast set of production processes, the capitalist system is simultaneously a polarizing system.

Polarization occurs through three tiers, labeled the core, semi-periphery, and periphery, where the category of "semi-periphery" allows for the possibility of mobility between core and periphery in the international system (Taylor 1985: 9–10).

Dependency and world systems theorists have criticized the World Bank on a number of grounds. Hayter (1981), for example, identifies three problems. First is the issue of accountability. He observes that the World Bank is organized in such a way that the voting rights of member nations are determined by their level of national financial support for the Bank. Because developing nations contribute far less, they have very little leverage over the policy directions of the Bank. Second, Hayter (ibid.: 88) notes that the "World Bank and IMF were set up after the Second World War to solve the problems of rich countries." Thus, the major purpose of the IMF and World Bank is to make the world safe and predictable for private capital and free trade. Finally, if countries refused to adhere to the policy recommendations and rules of the Bretton Woods institutions, these organizations could use their leverage to destroy the financial credibility of those obstinate countries. Thus, the World Bank and the IMF are said to perpetuate the asymmetry between the North and the South.

In her influential radical critique of the Bank, Payer (1982: 19) argues that the World Bank "promotes a philosophy of development to advance the interests of private, international capital in its expansion to every corner of the 'underdeveloped' world." Payer outlines the Bank's influence in developing countries' policies in the following manner. The first level of influence is that of individual projects. Payer argues that poor people seldom benefit from Bank projects, and that Bank initiatives help perpetuate the status quo. The second level of influence is at the sector level. Payer argues that when the Bank funds an individual project it demands changes in the entire sector impacted directly or indirectly by that specific project. For example, if the Bank lends for a mining project, it will often demand changes in legal and taxation codes concerning mining investment as well. Poverty often increases as a consequence of the Bank's intervention, according to Payer. Finally, the Bank is able to influence national policies by forcing borrowing governments to abandon progressive policies in favor of harsh austerity programs, which are often formulated in cooperation with the IMF and major Western powers. For Payer, the Bank

> is perhaps the most important instrument of the developed capitalist countries for prying state control of its Third World member countries out of the hands of nationalists and socialists who would regulate international capital's inroads, and turning power to the service of international capital.
>
> (ibid.: 20)

Theorists reflecting on the activities of the Bank through the dependency paradigm contend that the inherent internationalism of the Bank thwarts the Third World's aspirations for national sovereignty.

In sum, dependency and world systems theorists argue that the international division of labor is reproduced through the following mechanisms. First, the core countries impose a cultural hierarchy that promotes a prevailing consciousness throughout the world system. This includes variables such as media exchange and the export of an ideology that legitimates the core–periphery structure of the world system and inequalities inherent in the modern world economy. Second, there is direct organizational control over the economic and political processes of developing countries. Trade treaties and investment policies that favor the core are said to be important to the core's maintenance of its status at the apex of the hierarchy. Economic strength, military prowess, and political treaties all enable core countries to exert a considerable influence on the periphery and semi-periphery. The World Bank, then, is important in ensuring that the core retains leverage over domestic policies of states in the developing world. The Bank's structural adjustment programs are viewed as striking examples of predatory economic management of Third World economies by the Bretton Woods institutions and the developed core.

Dependency theorists argued that it would be in the best interest of Third World nations to break connections with international financial institutions like the World Bank. Beyond this recommendation, dependency theorists do not outline any major policy reforms for the Bank other than rhetorically advocating its demise or calling for its fundamental restructuring. However, dependency theory's forceful critique of Western imperialism, as well as its popularity during the 1960s, contributed indirectly toward pushing the Bank toward poverty-oriented lending for basic needs, such as shelter, as part of a geo-political calculation. This issue is discussed in the next chapter.

Postmodernism, development theory, and the World Bank

The "postmodern turn" in the social sciences and the humanities has also found its way into development studies. Like modernization and dependency thinkers, postmodernists also have attempted to theorize the relationship between global dynamics and local processes in developing countries. Post-modernists begin with the view that the current "new" period represents an epochal shift from the past in terms of style and method.

Postmodernism regards development as a discursive field, a system of power relations which produce what Foucault (1979: 12) calls the "domain of objects and rituals of truth." Using Foucault's theme of discursive power as well as the deconstructionist method of analyzing the representation of social reality, Escobar (1995) seeks to interrogate "development" in order to illustrate how the dominance of this system of knowledge has silenced non-Western knowledge systems, and how peasants, women, and nature are objectified and targeted by the "gaze of experts." His work has received considerable attention and is representative of the postmodernist critique.

The central premise of Escobar's book is that international discourse on development after World War II represents the exercise of power over the Third World, and that international development agencies such as the World Bank are instruments for achieving that aim. "Development," according to Escobar (ibid.: 13), "has relied exclusively on one knowledge system, the modern Western one." Focusing on three defining characteristics of the global development discourse – the process of knowledge production, which relates to and informs development; the wider power relations which shape development practice; and the types of subjectivity facilitated by development discourse – Escobar (ibid.: 13) observes that "most people in the West . . . have great difficulty thinking about Third World situations in terms other than those provided by the dominant development discourse."

Postmodernist development theory is also a response to modernization and dependency theories, both of which contain a modernist conceit that downplays traditional knowledges and cultures. Escobar's work is an attempt to steer development studies away from their preoccupations with neo-classical economics, on one hand, and political economy, on the other, toward issues of discourse, identity, and representation. Emphasizing the relationship between development and capitalist forces, Escobar points out that the socialist critique of development is incomplete. For Escobar, development economics presents itself as a science, making claims to objective and infallible truth, and its self-portrayal renders development discourse inimical to the traditions, realities, and aspirations of Third World communities.

Postmodern critics track the Western discourse of development from the creation of organizations such as the World Bank and the United Nations in the 1940s and 1950s, through the growth of a legion of "experts" to staff these institutions, as well as the successive strategies adopted by these agencies through the 1980s. They argue that people are left out of the discourse because of its elitist, ethnocentric, and technocratic method of reducing human beings to facts and figures. Postmodernists argue that, instead of solving the problems of poverty and hunger, development agencies have made them manageable while becoming the livelihood of an amorphous group of development "professionals." Thus, the postmodern project aims to deconstruct development, to expose the practices through which the discourse of development is reproduced, and to explore the alternatives that are available.

Escobar argues that expert planning through international development agencies is one of the principal practices in which the discourse of development is embedded, making it inherently incapable of addressing the world's development problems. Through its domination of development discourse as the largest development agency, the World Bank is able to construct a discourse in which all countries must participate. Using Foucault's ideas to describe the impact of the World Bank on developing countries, Escobar writes that:

> the impact of the World Bank goes well beyond the economic aspects. This institution should be seen as an agent of economic and cultural imperialism at the service of the global elite. As perhaps no other institution, the World Bank embodies the development apparatus. It deploys development with tremendous efficiency, establishing multiplicities in all corners of the Third World, from which the discourse extends and renews itself.
>
> (ibid.: 167)

The ubiquitous and all-powerful World Bank, in Escobar's view, applies paradigms drawn from advanced industrialized countries that have repeatedly proved to be insensitive to regional and local conditions in the developing world. His abstract account of how the Bank constructs the development discourse is based on his study of field representatives' framing of development problems, analysis of those problems, and, finally, their own representations of the problems. Bank-conceptualized "solutions" distort the world peasantry's problems, according to Escobar, but are still imposed over and above alternative or locally developed solutions.

In Escobar's view, development policy may have resulted in "forty years of incredibly irresponsible policies and programs" (ibid.: 217), but the World Bank "will not be driven out of business by repeated failure" (ibid.: 161). Drawing upon Foucault's observation that failure does not necessarily undermine social institutions, Escobar argues that development discourse incorporates new ideas and social movements, so that previously neglected groups and issues such as peasants, women, and the environment may be incorporated even as they are being marginalized.

Theories of development and the World Bank: a critique

Each of the schools of thought discussed above has influenced development thinking and outcomes in particular ways. However, they also contain critical conceptual flaws that lead to problems in their application. The ensuing discussion will explore some of the main conceptual weaknesses of the three approaches to development outlined above by situating them within the context of modernity. The section concludes that these schools of thought contain major problems that render them inadequate for understanding how the World Bank affects domestic policy choices in developing countries.

All three approaches to development discussed above, even the postmodernist, contain the metaphysical conceits of modernity. Modernity, a historic plexus of power, knowledge, and practice arguably originating in Europe in the sixteenth and seventeenth centuries, became, by the middle of the twentieth century, the dominant socio-cultural order of the world, adapting and extending its forms over time and space according to its universalizing logic. Wallerstein (1974) identified the emergence of a "modern world system" under modernity, centered on Europe with an attendant system of politico-economic structures and transactions organized around market exchange. As the central discourse of the so-called Enlightenment, modernity came to be culturally and philosophically associated with all that is "new," but also with a forward, rather than backward, gaze. Marx captured the turbulent and evanescent quality of modernity under capitalism with characteristic flair in *The Communist Manifesto* (Marx and Engels 1989: 12): "All fixed, fast-frozen relations, with their train of ancient and venerable prejudices, are swept away, all new-formed ones become antiquated before they can ossify. All that is solid melts into air." The cherished values of reason, rationality, and progress were all employed in the pursuit of what were identified as worthy goals: truth, beauty, and justice. Postmodernists and other critics of modernity are correct to question the modern agenda, but they frequently misidentify the problems of modernity as intrinsic to the goals and values of the Enlightenment when, in fact, they are probably attributable to the undemocratic way in which such goals and values were conceptualized and pursued.

Aesthetically and philosophically, modernity was concerned with the representation of the increasingly abstract, anonymous, and contingent quality of life in a world of rapid flux. Practically, modernity had appointed itself to the task of ridding human civilization of superstition, suffering, and tyranny. Modern assumptions color the way development is problematized by the major schools of thought as well as the way development is implemented in practice. These assumptions culminated in the modernization paradigm, which sought to make over the so-called "traditional" world in the image of the West. However, modernization cannot be separated from the dark side of European modernity, which confined human creativity and consciousness within Weber's famous "iron cage" of bureaucracy and control. Advances in science, technology, and culture – touted by the modernizationists as prescriptions for take-off in the developing world, but also surprisingly unquestioned in and of themselves by the dependency school – were accomplished through wars, various forms of violence, theft, slavery, and exploitation. Instrumentalist attitudes toward nature, the subordination of women, racism, and ethnocentrism were employed not only in the service of empire over the past two centuries, but also in the name of various revolutionary movements.

The modernization and dependency schools, both premised on the Cartesian rationalism inherent to modernity, accepted a world ordered by a series of hierarchical dualisms, even if they disagreed on which elements were superior in the hierarchy. For example, while the modernization paradigm was driven by a logic that privileged man, culture, future, science, and the modern over woman, nature, past, myth, and tradition, dependency theory confined itself to inverting the hierarchy of one particular dualism, core–periphery, while neglecting to examine other dualisms; furthermore, dependency theorists did not realize that the inversion of a hierarchical dualism still upholds it in principle. Hence, they reified the ubiquitous core, while modernizationists reified the periphery. As a result of their preoccupations with global and national problems, neither paradigm addressed the needs of ordinary people at the level of everyday life. Also, since both schools focused on factors external to the Third World, neither of them gave agency to the Third World. Modernization theory stressed the positive role of external actors, while dependency theory emphasized the negative effects of external economic and political linkages. The Third World, according to these views, either passively receives modernization through Western tutelage or is victimized by imperialism. Lipietz observes that:

> despite the undeniable formal superiority of the imperialism-dependency approach, it seems that, like the rival liberal approach [the stages of economic development], it had degenerated into an ahistoric dogmatism . . . if the South was stagnating, one theorist would tell you precisely what time it was, if new industrialization was taking place, another would say I told you so.
>
> (Lipietz 1987: 5–8)

Both the modernization and dependency schools are products of modernity, an "epoch that lives for the future, that opens itself up to the novelty of the future" precisely because it cannot appeal to myth or tradition for legitimacy (Habermas 1987a: 7).

However, Harvey (1989) notes that the intensity of time–space compression over the nineteenth and twentieth centuries, felt most acutely in Western and colonial cities, has led many commentators to declare a profound crisis of representation on the grounds that the modern epoch has ended. These observers prefer to speak instead of a condition of postmodernity, with an attendant style and method (Dear 1986). However, granted that they entail departures from the geography of modernity, these "new times" we are said to live in also curiously display several important continuities.

Researchers, like Escobar, who apply the postmodern method to development studies seek to discover the institutions, social processes, and economic relations on which the discursive formation of development is articulated. For them, discourses are power plays which assert a particular understanding through the construction of knowledge.

> Because they [discourses] organize reality in specific ways that involve particular epistemological claims, they provide legitimacy, and indeed provide the intellectual conditions for the possibility of particular institutional and political arrangements.
>
> (Dalby 1988: 416)

In doing so, discourses make these socially constructed arrangements appear natural, so as to "foreclose political possibilities and eliminate from consideration a multiplicity of words" (Dalby 1990: 4).

However, I contend that there is an air of déjà vu around many of Escobar's criticisms of the project of development. Echoing Horkheimer and Adorno's (1974) critique of the project of modernity, developmentalism is questioned for its evolutionary assumptions, its optimism about the possibility for solving global social problems through the expansion of production, its faith in modern science and technology, its reliance on experts, and its insensitivity to cultural diversity. In this regard, Escobar's critique, contrary to its own claims and intentions, reads like a modernist critique of a modernist project because it lacks originality in spite of an allegedly "postmodernist" point of departure. Ironically, the postmodernist approach does not acknowledge that it reproduces the very discourses of modernity it critiques when it conceptually totalizes (in spite of the postmodernist taboo against totalizing) and represents "development" as being deployed by the powerful West against the powerless Third World. Not surprisingly, therefore, the postmodernists' own totalizing account reveals modernist premises and yields modernist observations. For example, the modernist propensity toward determinism is evident in the postmodernist reification of development discourse as all-powerful in its capacity to organize the reality of the Third World. Like dependency theorists, postmodernists stress the negative consequences of the Third World's linkages with the West, thereby dualizing the relationship.

Methodologically, also, in spite of consciously distancing itself from "ideology," discourse analysis via deconstruction mimes and mimics the historical materialist theories of ideology that came of age with Marx, which were later interpreted through language as a medium of action by thinkers such as Thompson (1984) and Habermas (1987b). "Discourse" also resembles the idea of "hegemony" as developed by Gramsci (discussed in the next section). Echoing Gramsci, Escobar considers development as a "space" in which "only certain things could be said and even imagined" (Escobar 1995: 39). He then combines the articulation of ideas, institutions, practices, and changing historical realities into a unified discursive system. However, "discourse" does not capture the nuances of hegemony and ideology or their potential for praxis because it is, unlike hegemony and ideology, disconnected from critiques of other spheres of life, such as the economic.

Two important differences between discourse and ideology pertain to perspective and purpose. Postmodernists' analysis of development as a discourse is done with hindsight and supposedly without imputing any grand design or progress to past events that may inform the present. The backward gaze of the postmodern perspective draws upon genealogy to recount the past in order to know about the present. Applied to development, such a perspective implies that one cannot know the purpose of development as a discourse because the past is simply a random collage of events, and development outcomes in various countries are accidental. Postmodernism's critique of teleology also means that the future cannot be planned or intended. Such a conceptualization of time and purpose contradicts the experience of planning as a political practice and effectively locates power outside of the will of individuals and institutions. Foucault's (1978: 93) formulation of power, as a complex strategic relation that is everywhere, constitutes an abstraction of power from lived, experienced, practiced, and deeply scaled material reality. Said notes that:

Foucault takes a curiously passive and sterile view not so much of the uses of power, but how and why power is gained, used and held on to. This is the most dangerous consequence of his disagreement with Marxism, and its result is the least convincing aspect of his work. Even if one fully agrees with his view that what he calls the micro-physics of power is "exercised rather than possessed, that it is not the 'privilege', acquired or preserved, of the dominant class, but the overall effect of its strategic positions," the notions of class struggle and of class itself cannot therefore be reduced – along with the forcible taking of state power, economic domination, imperialist war, dependency relationships, resistances to power – to the status of superannuated nineteenth-century conceptions of political economy. However else power may be a kind of direct bureaucratic discipline and control, there are ascertainable changes stemming from who holds power and who dominates whom.

(Said 1983: 221)

Foucault's ideas, particularly his perspectives on the state and power, deserve attention here because they constitute the cornerstone of Escobar's thesis. The exercise of power has traditionally been conceptualized in terms of either the actions of individual or institutional agents or the effects of structures or systems. Weber, for example, conceptualizes the articulation of power relations as "systems of domination" and the "state bureaucracy." For Marx, power is rooted in the economic structure of society. Foucault's conception of power significantly departs from these views in that he calls for the close scrutiny of the "micro-physics of power relations" in different localities, contexts, and social situations (Harvey 1989). Such a shift led Foucault to conclude that there is an intimate relationship between systems of knowledge (discourses) that codify techniques and practices for the exercise of social control, on one hand, and domination within particular localized contexts, independent of any systematic strategy of class domination, on the other hand. The prison, the asylum, the hospital, and the university, for Foucault, are sites where dispersed and piecemeal organization of power is built up. However, what happens at each site cannot be understood by appealing to some overarching theory. For Said, Foucault lacks

something resembling Gramsci's analyses of hegemony, historical blocs, ensembles of relationship done from the perspective of an engaged political worker for whom the fascinated description of exercised power is never a substitute for trying to change power relationships within society.

(Said 1983: 222)

Because he draws so heavily from Foucault, Escobar's own ideas fail to empower the very "victims" of the discourse of development. Even as he claims to speak on their behalf, he strips them of whatever agency they might yet possess.

Postmodernists are certainly correct in questioning how development discourse homogenizes everything it encounters. However, characteristically, they discard the conceptual tool of generalization itself when countering a generalizing discourse. Their preoccupation with posturing against modernity has led them to philosophical commitments that do not permit them to selectively appreciate the analytically and politically strategic value of generalized representation, without which certain stories cannot be told. Such analytic tools are part of the arsenal with which authors and actors produce narrative scale as well as geo-political scale. Deconstructionists like Escobar circumvent such logical and practical impasses by conveniently shifting gears back to the modernist method of generalizing the discourse to suit their purposes. In spite of posturing to the contrary, they require generalization as a method in order to ascribe a cause and purpose to development (identified by Escobar as "domination"), without which they would have no reason to tell the story of development. However, by doing so, they take the bait of modernist reasoning in spite of their own theoretical prohibitions against it. Perhaps such missteps betray a modernist habit of mind that is unable to keep pace with the alleged postmodern rupture from a modern past.

Problems for praxis and policy

The conceptual issues I have outlined above, namely dualism, determinism, reification, and overgeneralization, pose numerous problems for practical application in general, but particularly for understanding the relationship between the World Bank and developing countries. Each of the major schools of thought discussed above has general prescriptions for Third World development. Modernization theory champions economic growth as the locomotive of development and promoted the export of Western technology and culture to help the Third World "take off." The dependency paradigm, preoccupied with external linkages, inevitably views delinking from the West as the solution to the Third World's problem of underdevelopment. Postmodernists, in turn, advocate the empowerment of local actors as an antidote to development. Some of the problems overlooked in their analyses include the diversity of development patterns in the Third World; the role of the state and other domestic actors; the balance between external influences and domestic politics; scale; and hegemony. I discuss each of these problems below.

The modernization and dependency paradigms, as postmodernists and other critics of the development process have pointed out, both fail to consider the historical and national diversity of development patterns that characterize the Third World. In the modernization paradigm, prespecified outcomes are associated with the development process, with little or no attempt to examine the spatial and temporal contingencies of the developing world. Wallerstein's (1976) world systems analysis, for example, collapses the formerly socialist states of Eastern Europe, mineral-exporting states such as South Africa, and newly industrializing countries of Latin America and East Asia as part of a homogeneous semi-periphery. While modernization theorists overlook the constraints imposed on developing countries by the global political economy, dependency theorists, on the other hand, "liquidate the unique history and development of specific countries" (Milkman 1979: 262). In this regard, Haggard (1990: 21) contends that in an "attempt to outline a parsimonious conception of international structure, [the dependency theorists] missed the variation in state strategies and capacities." As a result, neither modernization theory nor the dependency approach can explain differences in the behavior of similarly situated states. Postmodernists are, to their credit, interested in the particularities of places and peoples as they intersect with the "metanarrative" of development. Escobar (1995: 5) states that a focus on cultural issues has resulted in "new ways of thinking about representations of the Third World," but these ideas remain tentative and inadequately developed. When postmodernism's interest in representation, identity, and difference is stymied by its own reticence on economic and political issues concerning Third World actors, one is left with little more than good intentions. Postmodern thinkers on development often frame their analyses in excessively abstract language; their conjectures make little attempt to link symbolic and textual representations with political and economic representations. This is because they disregard an extensive body of theory on the materiality of capitalist development, exemplified by the works of Frank (1969), Emmanuel (1972), Amin (1974), Mandel (1980), and Smith (1984). The "post-development" alternatives suggested by postmodernists call for hybrid solutions that transcend tradition and modernity; however, postmodernists place the onus of finding these solutions upon less developed societies themselves, without sufficiently analyzing their capacity for autonomy in the context of deepening globalization.

As I argued earlier, the dualist underpinnings of the major schools of development obscure an elaborate choreography of external influences and domestic politics. As a result, the schools are forced to identify one as more important than the other. Kahler (1992: 10) argues that the proper measure of external influence on national policy is "the degree to which external actors change the trajectory of national policy from what it would have been in the absence of their intervention." He observes that the "slippage" between announced intentions of external actors and actual policy choices is substantial, suggesting the intervening weight of domestic politics. However, Kahler's skepticism about the actual leverage of international financial institutions such as the World Bank is not grounded solely on the role of domestic politics. Countervailing international factors can also undermine the influence of these organizations. In some instances, potential enforcers, such as creditor governments, have multiple, conflicting goals vis-à-vis the debtors. The concern to support a strategically important client can easily override the interest to enforce conditionality or even ensure repayment. Such arbitrary decisions can result in levels of external financing that are higher than what would have been provided otherwise. The Philippines under Marcos and Zaire under Mobutu are cases in point. While external pressures and influences do impact on national policy choices, political elites weigh the repercussions of external influences against their domestic political costs. What Kahler would call a "decision calculus" is insufficiently analyzed by the major development theories but is vital to the production of scale. This is discussed below and in the next section.

In examining the role of domestic politics, Haggard and Kaufman (1992: 15) recognize the relevance of international variables, which serve as "an important corrective, a reminder of the limits of analysis that rely exclusively on domestic variables." However, they identify three important limitations of perspectives that explain policy choice without referring to domestic political configurations. One problem is that longer-term sources of vulnerability to external pressure often lie in previous policy choices. Haggard and Kaufman hold that, although changes in commodity prices and terms of trade can be viewed as externally induced, the public investment booms that contributed to debt accumulation during the 1970s originated in domestic spending priorities and development strategies. These booms also contributed to overvalued exchange rates, which limited export growth and led to financial speculation and capital flight. Second, external shocks do not affect policy choice in an unambiguous way. Countries experiencing similar shocks adopted stabilization and adjustment programs at different paces with varying content. Korea, Chile, Costa Rica, and Ghana responded relatively early in the 1980s with comprehensive programs of stabilization and structural adjustment, while others such as Argentina, Peru, and Zambia did the opposite, and were not able to sustain structural adjustment policies of any sort. These variations suggest that domestic structures and choices are important in understanding national responses to external pressures.

The state is important as a mediating force between competing domestic and international interests. With respect to development, Skocpol (1977) has persuasively argued that development theories have not adequately conceptualized the role of the state. One main problem is that the state is viewed in largely instrumentalist terms because of these theories' inadequate grasp of politics. Dependency theorists, for example, see elites in Africa, and elsewhere in the Third World, as mere conduits or agents of foreign capital, lacking the capacity or will to enforce their own independent or national interests. Local elites are simply bearers of the global structures of dependency. Most theories, according to Skocpol (1987: 1080), have "managed to create a model that simultaneously gives a decisive role to international political domination and deprives politics of any independent

efficacy, reducing it to vulgar expressions of market-related interests."
Drawing upon her comparative study of revolutions, Skocpol (1979: 31)
concludes that states are not merely "analytic aspects of abstractly conceived
modes of production, or even political aspects of concrete class relations and
struggles . . . [they are] actually organizations controlling (or attempting to
control) territories and people." Therefore, for Skocpol, while states operate
in the context of an international system of states, they do have a logic and
purpose of their own. Following Skocpol, I would argue that any attempt to
describe the relationship between the World Bank and development must
understand the nature, role, and function of the state.

In light of deepening globalization, some arguments regard the state
as an archaic concept, with its influence and relevance waning relative to
supranational organizations such as the World Bank, the IMF, and non-
governmental organizations (NGOs) (see Knight 1989; Morss 1991; Strange
1996). The ever-increasing power of multinational corporations has resulted
in a deep pessimism about the ability of states to effectively manage
their traditional objectives in areas such as the promotion of economic
development and the regulation of economic activity. Glassman and Samatar
(1997: 164) point to a growing belief that states are "impotent in the face
of globalization." This sentiment is also expressed in, for example, Richard
O'Brien's (1992) provocatively titled book, *Global Financial Integration: The
End of Geography*. Stallings (1992) argues that it is premature to dispense
with the dependency paradigm's basic premises when examining the power
of international forces in determining policy options for the Third World.
She stresses the need to emphasize international factors when explaining
national policy choices, arguing that the primacy of the former impinges
on the latter in three distinct ways. First, international goods and capital
markets determine the availability of external resources, which, in turn,
sets important limits on the range of policy options. Second, policy choice is
influenced by transnational social and political networks that link domestic
and international actors. Finally, debtors are constrained by leverage, or the
financial, political, and ideological power exercised by creditors.

Just as they neglect the state, the major paradigms of development fail to
give sufficient attention to elites and local actors. On one hand, because they
privilege the role of external variables, neither the dependency paradigm
nor the modernization school sufficiently grasp how the general features of
the geo-political and economic system are inscribed into local structures.
Appadurai (1990) insightfully draws attention to how a variety of forces
which flow from particular centers of power are rapidly "indigenized,"
arguing that the influence of metropolitan ideas and models is of less concern
to many people than the power of local, regional, or national elites. Citing
the Indonesian experience, Appadurai illustrates that central to the Java-
centered elite's hegemonic project has been the production of a powerful
development discourse exhorting Indonesians to work together for national

development and economic take-off. Thus, for Appadurai, while the state-sponsored concept of development was obviously mediated by the broader global political economy, it is imperative to understand how global forces were appropriated and indigenized in order to legitimate the elite's socio-political order.

Postmodern analysis, on the other hand, emphasizes some local actors but neglects others, especially the state. Escobar's case studies, for example, focus on development professionals – planners, bureaucrats, and economists – as the group that is responsible for constructing the discourse of development, principally through their own pursuit of jobs and incomes. There is little analysis of local, national, economic, and military elites' roles in development; multinational corporations, the state, and other classes who have shaped development are also conspicuously absent. Escobar correctly critiques a development policy approach in which people are the passive recipients of modernity's ambiguous benefits. However, he proposes leaving the initiative to the local communities, even though he admits that they would be powerless in the absence of a supportive global framework. The existence of such a framework is, by his own account, unlikely. In this manner, postmodernists run aground with respect to policy because discourse analysis does not concern itself with how development is actually implemented. Development is seen as a "strategy without strategists" (Escobar 1995: 232, note 26). In his search for Foucault's "dispersion of power," Escobar overlooks the fact that power, accommodation, and resistance are deeply scaled. For this reason, I find that he is not able to establish, in spite of lengthy analysis, how the World Bank's strategy of integrated rural development (IRD) articulates with local, regional, and national configurations of power.

Scale and the production of space

Equally problematic is the fact that scale is often misconceptualized in theories of development. The "global" and the "local" are presented as two static, fixed, and separate spheres, with global sometimes corresponding to the West and local to the Third World. The major schools of development accept the conventional view of nested spatial hierarchies, in which the local is embedded within the national, which is then contained within the global. Spatial scale is reified to create discrete categories such as "global" and "local," which are then represented dualistically. Modernization theory, with its evangelism of Western ideas, culture, and technology, was part of a Cold War strategy that sought to "bridge" what was seen as a "gap" between the modern/global and traditional/local, in order to better integrate the Third World into Western capitalism. While dependency theory points out that the Third World can never develop as long as it is integrated into the capitalist world system, it responded with a "gap-ism" of its own. Because it also sees the global/core as separate from the local/periphery, it advocates "delinking" as a solution (Amin 1990). In this respect, postmodernists correctly evaluate external prescriptions for Third World development as totalizing, but they too fail to escape the global–local dualism in their advocacy of locally based, Third World resistance to global power. In sum, all these theories overlook the specificity of politics – global, national, or local – because of their dualistic interpretation of scale.

Following Beauregard (1995), one can identify two general tenets of "global–local thinking" in the major theories of development. First is the belief that global economic forces are the dominant forces and the starting point for any interaction between separate spatial scales. Global forces are seen as penetrating downward to the local scale, incorporating institutions, industries, people, and places at other scales, but bypassing them if they are resistant. Second, because theories of development conceptualize each spatial scale as independent of others, with diverse interests and resources, global forces are seen as mediated by adaptation or resistance at lower scales as they filter downward. Intermediate scales are simply seen as adding complexity to, but not altering, the global–local relationship. It is not surprising, then, that the major theories view states and non-governmental entities as either unimportant or reactive (whether resistant or compliant) to transnational forces, even as those entities construct legal, political, and financial infrastructures that enable capitalism to function.

After some critical scrutiny of development theories, I see a number of conceptual problems concerning not only the definition of spatial scale, but also its application. The global–local dualism inherent in theories of development reifies and ossifies spatial scale as an a priori phenomenon, which leads to a distorted understanding of the geographic concentration and reach of the power of actors. For instance, the dependency school characterizes and prioritizes the core's imposition of its economic and political will upon the periphery as domination; however, this is only a partial account of a complex power relationship. Similarly, postmodernists' (as well as dependency theorists') caricature of a "powerful," global, First World dominating the "powerless," local, Third World is the result of a serious underestimation of both the extent of Third World integration into the capitalist world economy as well as the ubiquity of "Western" culture. Theories of development that are critical of modernization typically fail to understand that the idea of domination holds only if there exists a conceptual separation and opposition between two entities in which the power of the dominant and the powerlessness of the subordinate are absolute. In this regard, it is useful to consider the ideas of Gramsci, who distinguished between "domination" in a state based upon force and "leadership" in a state based upon hegemonic consent. Gramsci wrote in his *Prison Notebooks* that

> The supremacy of a social group manifests itself in two ways, as "domination" and as "intellectual and moral leadership." A social group

> dominates antagonistic groups, which it tends to "liquidate," or to subjugate by armed force; it leads kindred and allied groups.
>
> (Gramsci 1971: 151)

In fact, "domination" does not apply to situations in which the "other" has been conceptually and practically incorporated into the "self." Because it assumes the relationship between advanced and developing countries to be one of domination rather than one of mutual but unequal need, development theory is unable to provide praxis-relevant alternatives to modernization and its agenda of the gradual, dependent incorporation of a consenting Third World.

The World Bank certainly has great political leverage over Third World nations, and developing countries are indeed part of a matrix of international, social, and institutional incentives and constraints that limit their range of policy alternatives. However, I find that the dependency and postmodern approaches to development fail to address how some countries circumvent international pressure, including that of the Bank, while others do not. In sum, these schools of thought are unable to grasp how power relations are scaled or how they articulate with the material and ideological power of transnational capital and the World Bank. I consider these issues in an alternative framework, below.

Conceptualizing the World Bank and development: an alternative framework

Left perspectives on development, such as the dependency and possibly the postmodern approaches, often regard the World Bank simply as an instrument of domination that impedes development and increases dependency. The role of the Bank is, in fact, far more complex. The modes of regulation that make up the world political economy are produced and reproduced through active, scaled human struggle, with the World Bank as a particular actor, albeit a powerful one. A more careful reading of the Bank's role in development is needed – one that avoids reducing hegemony to domination, and grasps the uneven and scaled articulation between the World Bank and developing countries. In this section, I shall present a synthesis of Gramsci and the French Regulation School, both of which are especially insightful for theorizing the role of the Bank in development. Examples from the experiences of several African countries are then presented to illustrate the complexity of the relationship between states, domestic and international actors, and international financial institutions.

Writing from the perspective of historical materialism, Gramsci nevertheless criticized it for being overdeterministic in predicting an inevitable change to a new mode of production, as if this change were preordained in history. Such teleological views of history led to what Gramsci called "fatalism," which is a resigned acceptance of whatever happens historically. He argued for a break with deterministic and mechanistic forms of Marxism, such as Althusser's structuralism, in which change is seen unproblematically as an inevitable outcome of the laws of history, working independently of the human will. Gramsci also avoids economic reductionism, also usually associated with historical materialism, through a dialectical conceptualization of base and superstructure, rather than a hierarchical and dualistic one. Theories of Third World development would do well to heed these warnings, in light of some of the criticisms I have raised in the previous section. In this respect, I believe that Gramsci's ideas are useful in addressing the problem of agency in development theory.

In Gramscian critical theory of international relations, the international as well as the local are "embedded in a space called hegemony. . . . hegemony [referring] to a historical fit between social forces, states and the world order" (Keyman 1997: 8). From this perspective, development is a zone of struggle, a manifestation of the hegemonic space produced by states, markets, and civil institutions. As a supranational financial institution, the World Bank shapes development by producing this hegemonic space unevenly through a "historic bloc" of actors with convergent interests, such as states or multilateral corporations. A historic bloc is a strategic alliance between a broad range of activities, values, norms, and practices that mark the multiple and complex foundations of the relationship in which structures and superstructures are joined in dynamic interdependence (Holub 1992). The concept of hegemony also retains the specificity and centrality of both the state and the interstate system without having to separate them analytically from civil society or the global political economy. Finally, hegemony grasps agency, but also recognizes structures without reifying them, as the world systems and dependency theories have often done.

According to Cox, hegemonic groups and institutions implement a universal language (norms and ideas) in an attempt to work together so that a multiplicity of interests are made compatible with one another:

> to become hegemonic, a state would have to found and protect a world order which was universal in conception, i.e. not an order in which one state directly exploits others, but an order which most other states (or at least those within reach of the hegemony) could find compatible with their interests. Such an order would hardly be conceived in inter-state terms alone, for this would likely bring to fore the opposition of state interests ... The hegemonic concept of world order is founded not only upon the reputation of inter-state conflict but also upon a globally conceived civil society, i.e., a mode of production of global extent which brings about links among social classes of the countries encompassed by it.
>
> (Cox 1993: 165)

Strategies of alliance between the World Bank, developing societies, states, elites, and capital must be viewed in light of the conditions in which those strategies unfold. Resistance, for Gramsci, entails a coalition of counter-hegemonic interests and a carefully planned strategy capable of tapping into the collective will and consciousness.

Gramsci conceptualized politics as articulation. Following Gramsci, Laclau and Mouffe (1985) argue that alliances among interest groups do not simply emerge from the mode of production, or some other component of a social formation; they have to be actively constructed, consciously articulated, and vigilantly maintained. For them, social, political, and economic "reality" is a construction made possible through "articulatory practices" which establish the identities of entities in relation to other entities. Therefore, for Laclau and Mouffe, all identity is relational. Articulation, then, may be conceptualized as the construction of "nodal points" that temporarily arrest meaning and identity from a state of "becoming" into a state of "being." Without nodal points, meaning and identity would perpetually slide away. Nodal points partially fix the meanings constructed by articulatory practices.

Human beings articulate with others at different times and in different places in a contradictory manner that is simultaneously enabling and disabling, unifying and conflicting. According to Marx, such contradictory articulations are the basis by which human beings make their history, under historically inherited conditions. For French Regulation theorist Lipietz (1986), unity is expressed in struggle, and it is precisely this struggle that introduces dynamism in time and space. The Regulation School is particularly concerned with explaining the coherence and continuity of everyday life in the face of conflict and chaos under capitalism. Every moment is a truce that keeps revolution at bay. Individuals and social entities alike negotiate and strike compromises for the continuity of a preferred way of life, uniting with allies in a struggle to forestall the radical transformation of their world by competing interests. Cooperation and coordination among actors produce stability and balance under capitalism, whose driving logic is competition and conflict.

Cooperation and competition in everyday life, according to Smith (1984, 1993; Swyngedouw 1997), are modulated by the compromise of scale, through which actors are able to contain conflict at particular sites of struggle or channel it fluidly to other sites as necessary. As sites of struggle, the seemingly distinct scales of the urban, the regional, the national, the supranational, and the global are simultaneously locations and relations of power that are constantly subject to change. For Smith, scale is the process by which opposing forces are temporarily reconciled, but also a strategy for moving conflict to a more manageable location. Homogenization is consented to at one place in order to resist differentiation at another, and vice versa, with scale deciding when and where. Similarly, empowerment and disempowerment are outcomes of scalar processes. Based on these premises I argue that scale is simultaneously a nodal point where meaning temporarily rests and also an articulatory practice through which meaning is created and can flow.

Social conflict is governed through a mode of regulation, which can be seen as an ensemble of practices or a set of rules embodied in institutions and forms of governance that assures the reproduction or transformation of relationships for capital accumulation and circulation to continue. Institutions may be forms of the state or other formal or informal configurations of governance, and part of a world configuration such as an international system of states (Lipietz 1987). Hegemony, then, is the capacity of a dominant group or a historic bloc to impose, through coercion or consent, a mode of regulation as a compromise desired by all the parties involved, at a particular spatial scale advantageous to the dominant group(s). Thus, a "geometry of power" (Massey 1992, 1993) is produced, maintained, and reconstituted unevenly through the interplay of hegemony and resistance. Geometries of power see stability and permanence, but they are periodically disrupted, partially or totally, by counter-hegemonic blocs. The process of "jumping scale" is a central strategy of actors, dominant or subordinate, for acquiring or increasing control in new geometries of power (Smith 1993). Seen in this light, uneven development is simultaneously an outcome of, and a strategy for, capital accumulation as enabled by these circumstances (Smith 1984).

The form of the state, which may be seen as an "ensemble of institutionalized compromises," is decisive in the organization of the mode of regulation (Delorme and André 1983, cited in Swyngedouw 1997). The state is also a crystallization of certain allied class interests in time and space that can act independently of, but also upon, civil society and the economy through its capacity for violence and law (Poulantzas 1978). The forms of the state and other institutions are similar, but not identical. Along with states, supranational institutions are uniquely positioned to organize modes of regulation, form new historic blocs, and produce scaled spaces in hegemonic ways. After World War II, a host of supranational organizations, such as the Bretton Woods institutions, the North Atlantic Treaty Organisation (NATO), and the Global Agreement on Tariffs and Trade (GATT), were established as part of a configuration of global governance. Since then, these organizations have embedded themselves within a series of scaled articulations in order to regulate compromises such as development, defense, and trade. The glocalization of international financial institutions, or their capability to act globally and locally, represents an upscaling of their hegemony in development, as argued by Swyngedouw (1992a,b) and Robertson (1995).

Therefore, I argue that the increasing power of institutions like the World Bank does not necessarily imply that the state is obsolete in the current socioeconomic and political order; instead, it ought to be conceptualized as a structure that is constantly undergoing a process of restructuring in order to meet the evolving needs of capitalism and imperialism. By no means should states be relegated to the "dustbins of history" (Glassman and Samatar

1997). While it is true that transnational organizations like the World Bank often exercise considerable influence over the domestic affairs of developing countries, the terms of the relationship are negotiated with domestic elites, who often exploit that relationship to promote their own interests.

Myrdal's "soft state" theory (Myrdal 1968) characterizes many African and other Third World states as "weak" or "soft" in terms of their institutional and policy implementation capabilities. Theorists of the soft state argue that dependence on external capitalist centers and/or the lack of systematic institutionalization prevent these states from authoritatively regulating the allocation of resources and power. However, I contend that such theories fail to grasp the reality of authoritarian, interventionist, and class-based regimes in the Third World, where bureaucratic control is highly centralized and power is consolidated in the hands of a small minority. Since independence, many Third World leaders, especially in Africa, have been relatively successful in protecting their control over domestic political systems against external pressures for change. Political elites continue to use state power to exercise great control over sectors of the economy that are major sources of income for the country (Shafer 1986, 1994). I argue that it is for this reason that supranational organizations seek articulation with Third World states. Over time, such articulations have resulted in a rescaling of the national state and the glocalization of the World Bank.

The World Bank's hegemonic production of space

Following Lefebvre (1991) and Poulantzas (1978), I conceptualize the World Bank as a powerful, supranational actor that actively participates, along with states and private capital, in the production of abstract space. Through hegemonic spatial practices such as housing, on one hand, and the representations of space produced by its officials (its economists, planners, and technocrats), on the other hand, the World Bank extends, in concert with the state, grids of power into the concrete spaces of developing countries. It is useful to briefly consider here Bank experts' accounts of what constitutes development, exemplified below by a debate over "basic needs." I regard this debate as an exercise in official representations of space that constitute the Bank's hegemony not only in development, but also in its articulation with Third World governments. Because housing is a "basic need," Bank-sponsored housing development emerges as a hegemonic spatial practice that is reinforced by the Bank's official statements and representations. The aim of this discussion is to show that the Bank's hegemony in development is derived from the manufacture of consent among actors at multiple scales. In this example, the Bank was able to generate consent in the form of a compromise at two levels: between the Bank and its critics on one hand and within the Bank itself, between opposing viewpoints on "basic needs," on the other. Representational and practical "compromises" are, thus, part of the Bank's scaled hegemonic strategy, as I shall describe below and in later chapters of the book.

During the 1970s, in a climate of mounting criticism that its large-scale infrastructure projects did not address social issues, the World Bank was pressed to preserve its relevance and legitimacy as a development organization. This issue is discussed at length in the next chapter, but I note here that in response to such pressure the Bank began to adopt a classic strategy known as "semantic infiltration," the analysis of which provides much insight into language as ideology and hegemonic practice. The term "semantic infiltration" was coined by American Senator D.P. Moynihan (cited in Steinberg 1995), who advised President Johnson to employ the tactic in a 1965 speech as a means of defusing mounting tensions over civil rights and poverty. As old as the history of diplomacy itself, the technique refers to the appropriation of the language and posture of one's opponents for the purpose of blurring political distinctions between them and oneself. It is my observation that the World Bank often mimics the language of its critics through a rhetorical sleight of hand as it attempts to move key debates in politically safe directions, thereby neutralizing counter-claims by adopting some of them for itself. Robert McNamara's poverty reduction programs may be seen in this light: the Bank absorbed the language of the dependency theorists' critique of the growing inequality of less developed countries, and then synchronized this appropriated lexicon with its own agenda. For example, McNamara himself stated in his famous 1973 "Nairobi address" that "the basic problem with poverty is that growth is not equitably reaching the poor" (McNamara 1981: 237).

The Bank's ultimate mimesis of the "basic needs" school reveals another aspect of semantic infiltration: in the ability to defuse unfavorable ideas, opinions, or recommendations. The "basic needs" approach, popularized by Seers (1969), gave greater importance to "need" as a basis for allocating resources, as opposed to "output." Proponents of the basic needs approach emphasized fundamental human needs, such as nutrition and health, but also nonmaterial needs such as human rights, self-determination, self-reliance, political freedom, security, participation in decision-making, cultural and national identity, as well as a sense of purpose in life and work. Within the Bank, the "basic needs" idea as an ethical proposition was doomed to fail from the start, given conservatives' opposition to it at the time. In spite of efforts by Mahbub ul Haq, senior policy advisor to Robert McNamara, and others to push the Bank toward poverty-based lending (discussed in the next chapter), the issue of "basic needs" was always met with doubt as to whether fulfilling them was possible or even desirable (Kapur et al. 1997). Of major concern for the conservative camp was the fact that greater government intervention in both production and distribution would inevitably result as a corollary of the basic needs approach. All sides of the debate were encouraged by McNamara, who sought to determine how policy would finally crystallize, but his seemingly open stance had the effect of frustrating the basic needs proposal during the mid-1970s. In due course, great semantic uncertainty emerged, given the wide range of subjective interpretations of the words "basic" and "needs."

McNamara himself contributed to the overall ambiguity surrounding the trade-off between "growth" and "need" by insisting that he wanted both. This had the effect of terminating the discussion for some time.

By the late 1970s, the "basic needs" idea had regained currency in some circles within the Bank, who had a degree of tolerance and optimism for it, given the political climate of the times, but it was still ill-received by the majority of Bank staff, who felt that such "statist" policies were not becoming of the Bank. As a result, proponents of basic needs were pressured to justify their proposals in terms of cost and economic return, and to assume the burden of proving that economic growth would not be impaired. Conservatives insisted that there was a trade-off between growth and meeting need, and that ethical claims were not admissible criteria for Bank lending. In the end, as a response to internal and external pressure, the World Bank moved to adopt the "basic needs" concept, but with a crucial omission: the Bank's final word on the debate, as put forth in the book written by Streeten *et al.* (1981), made no mention of nonmaterial needs.

Thus, in spite of being late to enter the basic needs debate, when the Bank finally finished debating, it had arrived at a "compromise for everyone" that actually suited its own agenda best. In doing so, the Bank was able to effectively arrest the meaning of "basic needs" and conclude the debate with its own definition. By the time McNamara left, the World Bank had successfully neutralized or deflected, in a similar fashion, some of the harsher criticisms from the Left, while keeping in tune with its hard-liners. At any rate, "basic needs" soon became passé as the Bank's rhetoric began to settle more comfortably into "structural adjustment."

It is interesting to note that the Bank, in much the same way it toyed with the "basic needs" concept, is currently placating critics with a new motto, "Our dream is a poverty-free world." It is premature to determine whether the Bank is serious about poverty, or whether, as with basic needs, it will discard the slogan altogether after it has ceased to be politically useful. However, the Bank's appropriation of key words in development, such as "gender" and "the environment" are certainly examples of how the Bank appropriates rhetorical and semantic spaces, and how it routinely deploys language as a hegemonic practice. I have cited these examples in order to illustrate that the act of compromise or agreement represents a temporary, strategic gelling of viewpoints that capture the historic moment. This compromise is then coordinated with other similarly constituted compromises at other locations and times, thereby creating channels of consent through which policy may then be formulated, disseminated, and implemented. Scale is actively produced in the process as a solution to contemporary problems, but according to the relative power of competing actors. I am not suggesting that the Bank's use of terms such as "basic needs" is wholly hypocritical but, rather, that shifts in the Bank's policies correspond to the prevailing political climate, both internally and externally. The fact that ideas such as "basic needs" had even a brief career at the Bank means that the Bank does not implement its policies through coercion alone, as commentators have often argued.

Development, the World Bank, and Third World states

In the present political configuration of the capitalist system, I argue that the World Bank's hegemony entails its partial leadership, not overt domination, within a historic bloc of supranational organizations, states, and multinational corporations, as well as international and domestic elites. From a Gramscian perspective, the Bank's hegemonic leadership in Third World development is partly a function of its articulation with national states. That articulation, in turn, hinges upon the integrity of the national state as a geopolitical actor. The hegemony of the national state varies between complete integration within its social formation, at one end of the spectrum, and an almost total lack of organization at the other. Under full integration, relatively stable states with domestic hegemonic leadership consent to the World Bank's leadership, whereas less stable states suffering from crises of legitimacy are vulnerable to coercion by the Bank. In the following discussion, I shall describe how such an articulation, of which the politics and representation of development are the outcomes, sets the stage for the World Bank's relationship with developing countries.

By recognizing that the economic alternatives of Third World countries are severely constrained by their dependent integration into the world economy, I do not conclude that these governments have no options, or that their dependence on foreign powers and sources of finance is absolute. In reality, domestic policies do matter. Dependency, I argue, creates an environment of constrained opportunity, not always complete domestic impairment. Elites holding on to power in some Third World states often have an interest in maintaining the very conditions of crisis. These conditions often lead to multiple renegotiations of debt, for instance, and increases in humanitarian assistance and development aid, which then become windfalls of wealth for corrupt officials. According to Fatton (1989: 182), there is some irony in that "the incompetence of some Third World states in managing the process of development is an incompetence that rewards the incompetent themselves." The World Bank has recognized such "incompetence" as an opportunity to enter into the domestic affairs of many developing countries. The political economy of debt servicing has forced many countries into a series of foreign-induced "structural adjustments." Given their need for assistance in repaying debt, these nations have had to acquiesce to programs designed by the World Bank which entail, above all, a process of economic privatization that seeks to streamline government bureaucracies, curb state control, reduce public expenditure, and "free" the market. The ultimate goal of these structural adjustment programs is to maximize the withdrawal of the state from the economy (Kolko 1988).

The intrusion of foreign institutions, such as the World Bank, into the domestic policies of Third World countries has raised the specter of a new colonialism. However, it is necessary to bear in mind that neocolonialism exists because it serves the material, political, and strategic interests of not only external, capitalist, geo-political forces, but also those of the domestic ruling classes in many Third World countries. In addition to the influence of outside agents, it is through a domestic apparatus of repression that indigenous elites organize their dominance. Thus, as Fatton notes with respect to Africa, while external dependence is real, extensive, and constraining, it is neither absolute nor unilateral: "the ruling classes . . . have demonstrated Machiavellian imagination and statecraft in maneuvering the terms and conditions of dependence to their own corporate advantage" (Fatton 1989: 183). Thus, I would argue that, as dependence is articulated by internal constellations of power and class, its effects are contradictory and scaled, with differing experiences for various classes. In other words, the closer one is to state power, the more one benefits from dependence; as a corollary, the farther one is from state power, as are the poor, the greater the losses stemming from dependent articulation and integration.

This observation is useful in explaining the "bread riots" or protests that erupted in many parts of the Third World after the imposition of World Bank and IMF-recommended austerity measures (Walton 1987). According to Shafer, these protests ought to be understood in terms of "proximate *vs.* deep cause."[2] While the proximate cause of the riots in a particular country may have been the IMF stabilization program, Shafer (1994) points out that the program is in place only because the country is already experiencing economic problems. Therefore, the real issue for Shafer (ibid.: 34) is, "Is the country bankrupt because its policy options and actual policy course were so distorted by international pressures, or because the government's trade, investment, monetary and fiscal policies produced an untenable situation that necessitated IMF and World Bank intervention as part of a desperate effort to control the problem?" For example, the Zambian government, according to Shafer, cut food subsidies (which accounted for about 35 percent of the national budget at the time) in the early 1980s at the behest of the IMF. When riots followed, President Kaunda blamed the IMF and World Bank. However, Zambia's agricultural sector and national food production had been systematically degraded as a matter of national policy all along, according to Shafer. Although Zambia had exported food in 1964, by the time of the riots it had become heavily dependent on South Africa and Zimbabwe for imported foodstuffs.

While there are many important and valid criticisms against IMF and World Bank policies, it is important to place them in context. Few countries approach the IMF and World Bank pro-actively, on their own terms, before a crisis erupts, which suggests that there may be other facets to the accounts of imposed stabilization measures. According to Shafer, IMF and World Bank policies often have adverse consequences for the impoverished masses of the Third World because dominant upper classes are able to shift the burden of stabilization onto the poor. In Zambia, Shafer (ibid.) argues, the depth of food subsidy cuts was partly the result of higher-level bureaucrats' ability to resist pay freezes. The specific policy choices of governments and their concomitant impact on national development agendas are not entirely blameless for the development outcomes in many Third World countries. Third World development takes place within the context of inherited structures of inequality and imperialism, to be sure, but it is also influenced by domestic politics. As Ravenhill (1986: 3) observes, "it is entirely incorrect to suggest that governments enjoy no autonomy from international forces." Domestic elites made many development choices which have had adverse consequences and for which they must be held accountable. For example, many African countries failed to consider local and/or more efficient options before adopting import substitution (Kincaid and Portes 1994). Nationalization, another strategy adopted by many Third World governments, often had a highly selective impact on the poor, as Moll (1988) has shown in Latin America.

In his analysis of the Zairian debt, Callaghy (1986) has also demonstrated that external influence can explain domestic policy choices only to a certain extent. Dependence generates important benefits for certain classes:

> Mobutu goes to great lengths not to repay his debts, except with new debts. Borrowing and non-payment of debts are the central feature of the Zairian state. Mobutu knows that lending is a two way street, and he has shrewdly played the debt repayment game by attempting to manipulate slightly shifting coalitions of external actors, and the financial, economic, and politico-strategic interests they seek to protect and expand ... Mobutu is managing his dependence for survival, however, not for economic development or the welfare of the mass of Zairians. Given the severity of Zaire's situation, Mobutu and his political aristocracy have done amazingly well in this game of brinkmanship. They may not understand the finer technicalities of the financial system, but they do understand the politics of international finance ... the Zairian political aristocracy has adroitly blocked all efforts by international lenders to control its financial practice. Their record on this point is very clear.
>
> (ibid.: 317)

The Zairian state's capacity to resist what it perceived to be disadvantageous aspects of structural adjustment – that is, the restoration of fiscal order and the curbing of massive public corruption – symbolizes not the "softness" of the former Mobutu regime, but the hard class interests of the political aristocracy. In this case, dependence rewarded public corruption and enhanced the privilege of the ruling classes until the regime collapsed in 1997.

In post-independence Nigeria, an extraordinary level of corruption was generated by the oil boom, with members of the "rentier state" taking control of the petroleum industry in the context of a weak and fragmented bourgeoisie (Watts 1983). Watts observes that petroleum-inspired accumulation and state expansion did not generate a national bourgeoisie capable of seizing control of the Nigerian state. Instead, the state took over large segments of the economy, often well beyond its managerial capacities. The Nigerian state is still under the control of a fragmented dominant class and state managers who have little capacity to foster systematic accumulation. In his 1983 study, Watts analyzes the Nigerian state's hijacking of the development process through a "triple alliance" between foreign, local, and state capital. He argues that the oil boom spurred a concentration of power within the state, elevating a few state managers to positions of considerable influence while marginalizing the rest of the population.

According to Simon (1992), the current IMF and World Bank structural adjustment and economic recovery programs (discussed in Chapter 4) should not be interpreted simply as attempts to promote market-led trade; they also "attest to their [the Bank's and IMF's] perception of large scale state structures and bureaucracies as key contributors to the current plight of most Third World countries" (ibid.: 9). A crucial element in the conventional wisdom of IMF and World Bank-sponsored structural adjustment and economic recovery programs is the reduction in the role of the state. Given the corruption of the ruling elites mentioned earlier, recognition of the need for political and institutional reform is widely shared in the Third World, even by individuals who are not sympathetic to the agendas of the IMF and the World Bank (ul Haq 1998).

However, in characteristic fashion, the World Bank has employed semantic infiltration to hijack the popular call for accountable government, greater representation, and political reform. Cloaked in the Bank's promotion of "governance" as part of its development agenda is the familiar call for a reduced role of the national state in the domestic economy. "Governance," like "structural adjustment" before it, poses another opportunity for hegemonic articulation with national states. In this light, it is possible to interpret the Bank's entry into governance as indicative of a weakening national state in a climate of globalization. The question, I argue, is not whether globalization has rendered the state completely impotent and supranational organizations like the World Bank omnipotent, but how a rescaling of the state corresponds to an escalation of the World Bank's power.

The national states of developing countries have varying capacities to regulate the contestations of newly evolving socioeconomic, political, and spatial actors and practices at different scales. In such a conceptualization, I see national states as playing decisive roles, not only in the inflection of hegemony, but also in the expression of resistance to emerging power geometries. For instance, there is considerable variation in how developing countries accepted austerity, structural adjustment, fiscal management, governance, etc. While some states concurred with the prescription package, others protested and won concessions. Popular protests, such as the "bread riots," were often used by states as leverage to renegotiate debt or terms of loans, for example. Resistance movements against the World Bank and IMF policy, in turn, gained legitimacy and clout when they were allied with the state. Such observations only serve to highlight the importance of states and their articulation with civil societies.

The ascending neo-liberal orthodoxy in the study of globalization argues that the disintegration of national economies and the irrelevance of the national state have resulted in an economic and political landscape that *New York Times* columnist Thomas Friedman (2005) calls a "flat world." In contrast, I argue that the scaled relationships that produce highly uneven topographies of power in a long-globalizing world are more complex than Friedman's simplistic metaphor is able to convey. In fact, it seems absurd to conclude with Friedman that the might of the American state is now being leveled by globalization to match that of Zimbabwe. In reality, while the power of the nation state may be declining in some ways, this process is certainly not even; in fact, power may be consolidating in some places (Smith 2002).

The impact of the World Bank's policies on the developing world has been the subject of much debate and discussion in development studies. Some have argued that developing countries are held hostage by World Bank policy and are powerless to structure alternative courses of development for themselves. Others have criticized such a view, arguing against too much emphasis on external agents such as the World Bank, which would make developing countries the objects rather than the subjects of their own histories. Instead of privileging the "local" or the "global," or even framing the discussion in terms of such a dichotomy, this chapter has argued that the World Bank's role in development ought to be understood relationally in terms of the production of space and scale. These theoretical perspectives inform the book's consideration of the World Bank's role in urban development.

Chapter 2

Toward social lending

Shifts in the World Bank's development thinking

During the first two decades of its existence, the World Bank had a very narrow focus within Third World development, with most of its funding directed toward transportation and public utility projects, especially electricity. By the late 1960s, however, in response to internal and external forces, the Bank's development programs became more socially oriented. The aim of this chapter is to identify the trends and transitions in the Bank's development philosophy in order to understand how and why the institution embraced urban lending in the 1970s. This chapter is divided into three parts. The first section discusses the development philosophy behind the World Bank's early loan programs. The second part examines how multilateral and bilateral development institutions, and eventually the World Bank, came to recognize social concerns in developing countries. This reorientation influenced the Bank to re-examine its hands-off attitude toward funding urban development programs. Finally, Robert McNamara's presidency (1968–81) is discussed in some detail, as those years deeply influenced the Bank's thought and action pertaining to urban issues; the Bank's urban programs emerged from the initiatives undertaken during the McNamara presidency.

The World Bank's early development philosophy

Officially named the International Bank for Reconstruction and Development (IBRD), the World Bank is one of the two Bretton Woods institutions (the other is the IMF) established, first, to assist in the reconstruction and restoration of war-ravaged Europe and, later, to support the development process of LDCs. John Maynard Keynes, who chaired the commission that drafted the Bank's Articles of Agreement at Bretton Woods, outlined the following priorities for the Bank:

> It is likely, in my judgment, that the field of reconstruction from the consequences of the war will mainly occupy the proposed Bank in its early days. But, as soon as possible, and with increasing evidence as time goes on, there is a second primary duty laid upon it, namely, to develop

the resources and productive capacity of the world, with special attention to the less developed countries.[1]

The fact that Keynes prioritized European reconstruction is also noted by William Clark[2] (1986: xi), who questioned Keynes about how the reconstruction demands of Europe were to be balanced with the development needs of poorer countries. The developing world "must wait until the reconstruction of Europe is much further advanced," Keynes replied, adding cautiously that, in time, the Bank could assist in the creation of "a brave new world without beggars, even in Calcutta" (ibid.: xi).

During the Bank's early meetings, Third World delegates[3] were concerned that their countries' specific developmental needs would be ignored, that developed nations would not be differentiated from developing ones, and that no distinctions would be made between developed economies in need of reconstruction aid and less developed countries in need of development aid. These delegates urged participants to give priority to development, and not just reconstruction (Singer 1976). For instance, both Venezuela and Mexico submitted amendments to put development first, or at least "on the same footing as European reconstruction." Mexico argued that the Bank was meant to outlive reconstruction, and that the development of Third World nations would create the output of raw materials and the markets that Europe needed. The Mexican delegation also suggested that the Bank's priorities be reversed:

> What we ask is only that . . . in the event that countries requiring loans for developmental purposes do not use up the resources and facilities made available to them, countries requiring loans for reconstruction projects could have a claim on the unused funds.[4]

A forceful effort on behalf of developing countries also arose in the commission charged with drafting the Charter for the IMF. Sir Shanmukham Chetty, a member of the Indian delegation, stated that the Fund's Charter should "mention specifically the needs of economically backward countries." Although Chetty's proposal was not accepted at the Fund, it did make its way into the Bank's Charter. Originally, the Bank's Charter stated that the institution was "to assist . . . member countries," without distinction. However, as a result of pressure from Third World delegates at Bretton Woods, especially India, Article I(i) of the Charter came to read:

> The purposes of the Bank are: to assist in the reconstruction and development of territories of members . . . including the restoration of economies destroyed or disrupted by war . . . and the encouragement of the development of productive facilities and resources in less developed countries.

The Bank's first four loans all went to European states. The first loan, made one year after the Bank's opening, amounted to US$250 million and went to France for reconstruction. In the following year and half, loans were designated almost entirely for post-war reconstruction in the Netherlands, Luxembourg, and Denmark.

However, the World Bank's role in the reconstruction of Europe was soon cut short by the implementation of the Marshall Plan, which involved an intensive bilateral program to rehabilitate and integrate Europe. The United States government felt that any massive aid program to stabilize and reconstruct Europe should be under direct control of the United States, not administered through international development agencies. US Secretary of State Dean Acheson first alluded to this idea in a speech in Mississippi on May 8, 1947:

> It is generally agreed that until the various countries of the world get on their feet and become self-supporting there can be no political and economic stability in the world or no lasting peace and prosperity for any of us. Without outside aid, the process of recovery in many countries would take so long as to give rise to hopelessness and despair ... Requests for further United States aid may reach us through the International Bank or through the Export–Import Bank, or they may be of a type which existing national and international institutions are not equipped to handle and therefore may be made directly through diplomatic channels.
>
> (cited in Gardner 1969: 301–2)

Soon thereafter, on June 5, 1947, Under-Secretary of State George C. Marshall delivered his famous address at Harvard University calling for a comprehensive bilateral program for the reconstruction of Europe:

> The truth of the matter is that Europe's requirements for the next three years for foreign food and other essential products – principally from America – are so much greater than her present ability to pay that she must have substantial additional help, or face economic, social and political deterioration of a very grave character ... The United States should do whatever it is able to do to assist in the return of normal economic health in the world, without which, there can be no political stability and no assured peace ... Such assistance, I am convinced, must not be on a piecemeal basis as various crises develop. Any assistance that this government may render in the future should provide a cure rather than be merely palliative.[5]

As the Marshall Plan gained momentum, the Bank was forced out of recon-struction and began to focus increasingly on assisting LDCs as a way of en-suring its own survival and relevance. As Gardner observed,

The new measures devised to deal with the post-war disequilibrium soon overshadowed the financial institutions designed in the war. The normal objectives of the International Monetary Fund were gradually subordinated to the immediate requirements of European recovery ... The operations of the International Bank also yielded priority to the new program of reconstruction aid, but, in practice, the Western European countries preferred Marshall Aid to the assistance from the Bank, since the former came as grants or loans on easier credit terms than the Bank could make available. Therefore, as the Marshall Plan gained momentum, the Bank moved out of the reconstruction field. It turned instead, somewhat modestly at first, to the job of helping under-developed countries.

(Gardner 1969: 303–4)

The first development loan from the Bank was made to Chile in March of 1948 (Table 2.1), marking the Bank's first official entry into the uncharted territory of lending to LDCs. As Davidson Sommers, vice-president of the World Bank until 1959, recalls, "when the Bank started, people in the West didn't know much about the developing world except as colonies. They didn't know much about development lending, didn't know much about development economics."[6] The Bank's emerging concern with development-related issues in the Third World was reflected in its 1949 annual report, titled *The Role of the Bank: Economic Development,* which analyzed "the general conditions of poverty in the underdeveloped areas and their causes":

difficulties arise from the social structure of many of the underdeveloped nations where there are extremes of wealth and poverty. In such cases, strong vested interests often resist any changes that would alter their position. In particular, the maintenance in a number of countries of inefficient and oppressive systems of land tenure militates against increase in agricultural output and improvements in the general standard of living.[7]

(IBRD 1949: 47)

Senior Bank officials subsequently visited eighteen Latin American countries and eight countries in other parts of the developing world to determine how the Bank could best assist Third World development.

During the 1950s, the Bank also cautiously began to shift its role from that of a "bank" to that of a "development bank." This change in emphasis was highlighted in Bank president Eugene Black's 1956 speech at the annual general meeting. Calling the Bank a "development agency," Black stated that, "though originally conceived solely as a financial institution, the Bank has evolved into a development agency which uses its financial resources as a means of helping its members."[8] However, there was still much uncertainty and debate within the Bank over how much should be lent for Third World

Table 2.1 / World Bank loans as of June 30, 1952, by purpose and area, in millions of US dollars

Purpose	Africa	Asia and Middle East	Australia	Europe	Western hemisphere	Total
Reconstruction loans[a]	–	–	–	497	–	497
Other loans, total	125	129	100	202	329	885
Electric power[b]	58	19	27	34	253	391
Transportation						
Railroads[c]	18	63	14	1	12	108
Shipping[d]	–	–	–	12	–	12
Airlines[e]	–	–	–	7	–	7
Roads[f]	5	–	6	–	20	31
Ports[g]	1	4	–	13	3	21
Subtotal	24	67	20	33	35	179
Communication[h]	1	–	–	–	24	25
Agriculture and forestry						
Mechanization[i]	–	–	29	2	12	43
Irrigation and flood control[j]	–	31	10	13	1	55
Land improvement[k]	–	12	5	1	2	20
Grain storage[l]	–	–	–	4	1	5
Timber production[m]	–	–	–	5	–	5
Subtotal	–	43	44	25	16	128

Industry						
Manufacturing machinery	—	—	6	53	—	59
Mining equipment	—	—	3	8	—	11
Subtotal	—	—	9	61	—	70
General development						
Development banks	2	—	—	9	1	12
General development plans	40	—	—	40	—	80
Subtotal	42	—	—	49	1	92
Total, all loans	125	129	100	699	329	1382

Source: IBRD, *Annual Report* 1951.

Notes
a To France, the Netherlands, Denmark, and Luxembourg.
b For machinery, equipment, and construction materials.
c For locomotives, rolling stock, rails, and shop supplies.
d For vessels and marine equipment.
e For planes and spare parts.
f For building machinery and equipment.
g For docks, loading and dredging machinery, and harbor craft.
h For telephone and telegraph equipment and supplies.
l For general farm machinery and equipment.
j For construction equipment and materials.
k For machinery, equipment, and construction materials.
l For construction materials.
m For machinery and vehicles.

development, which countries and regions should receive development assistance, and which economic sectors should be supported.

In light of the Marshall Plan's assumption of responsibility for European reconstruction, on one hand, and the Third World's growing demand for development aid on the other hand, the Bank realized that its own future lay in assisting developing nations. However, it was cautious about how its lending programs would be structured. As most of the Bank's money was raised in the international financial markets,[9] the Bank was influenced by its own need to maintain good relations with a skeptical, post-Depression Wall Street. Among the investment community, many were unfamiliar with the Third World, regarded it as a great liability, and saw it as embroiled in nationalist and Soviet-inspired rebellions. To maintain its close contact with the financial community and convince them of its need to intervene in the developing world, the Bank frequently held conferences with investors at its headquarters in Washington, and established a marketing department in New York, which remained in place until 1963. The Bank sought to be actively engaged in Third World development without alienating its financial backers from Wall Street, arguing that its support of prudently chosen development projects would enable it to meet both objectives. When asked whether the New York Stock Exchange would react adversely to the Bank's financing of education, public health, and housing, Robert Cavanaugh, chief fund-raiser at the Bank from 1947 to 1959, replied:

> If we got into the social field . . . then the bond market would definitely feel that we were not acting prudently from a financial standpoint . . . If you start financing schools and hospitals and other water works, and so forth, these things don't normally and directly increase the ability of a country to repay its loans.[10]

According to Mason and Asher (1973: 110), the Bank feared that financing social development and poverty reduction projects "might open the door to vastly increased demands for loans and raise the hackles anew in Wall Street about the 'soundness' of the Bank's management." Thus, during the first post-war decade, Bank management felt that eliminating malaria, reducing illiteracy, building vocational schools, establishing clinics, etc., were unsuitable projects for financing by the Bank.

According to its Articles of Agreement, the Bank was expected to finance only "productive" projects for which other financing was not available on reasonable terms. In attempting to find appropriate projects, the Bank's management argued that, although costly equipment from abroad would be required for electric power plants, transportation, and communication systems, such infrastructure projects were, nevertheless, vital to attracting foreign investment to the developing world, especially as such projects would not be undertaken by the private sector. Thus, the Bank concluded that projects to develop electric power and transportation facilities were especially appropriate for Bank financing, as they would act as investment multipliers. It must be noted here that such thinking reveals the Bank's interest in keeping its eye on the financial markets. As a consequence, from the 1950s to the mid-1960s, the Bank placed an overwhelming emphasis on infrastructure-related projects. A review of the Bank's activities by sector during this period indicates that transportation, electric power, and industry constituted its main lending activities and dominated its loan portfolio (Table 2.2).

Only a small fraction of funds was made available for agriculture, and no funding was allocated for education, health, or other "social" needs. While the Bank did recognize that investments of many kinds were needed for development, it argued that certain kinds were more essential than others. As Mason and Asher note,

> The relative ease with which [the Bank] could finance electric power, transportation, and economic infrastructure projects ... made it an exponent of the thesis that public utility projects, accompanied by financial stability and the encouragement of private investment, could do more than almost anything to trigger development.
>
> (ibid.: 151–2)

The Bank thus came to embrace the view that investment in transportation and communication facilities, port developments, power projects, and other public utilities was a precondition for developing the rest of the economy.

In fact, the Bank was, at first, unyielding in its commitment to power and transportation projects, as noted by Lauchlin Currie (1981), a former special advisor to US president Franklin Roosevelt who led a general survey mission to Colombia in 1949 in order to prepare the country for a World Bank loan. While in Colombia, Currie accepted, privately, an invitation by the Colombian government to advise a committee to study the reorganization of the government. Currie then used this opportunity to seek the advice of this committee for his World Bank report, believing that "having the recommendations studied by a prestigious group of civic-minded citizens, ... top Bank officials would not need to feel they were responsible for recommendations in the fields of education, health and public administration" (ibid.: 61). Currie then persuaded Bank vice-president Robert Garner[11] to visit Colombia in order to discuss the Bank's report. Having presented Garner with a program of wide-ranging reforms consisting not only of the usual roads and energy projects but also, contrary to usual practice, ones for schools and public health, Currie recalls Garner's reaction: "One fateful day, Garner suddenly realized where I was leading him and drew back, saying, 'Damn it Lauch, we can't go around messing with education and health. We're a bank!'" (ibid.: 61–2). Currie concluded that the Bank "missed an opportunity to establish a precedent of linking non-bankable with bankable projects in an overall country program" (ibid.). A similar Bank reaction is evident in the case of Nicaragua, where the Bank ignored the recommendations of the 1952 survey mission's report, which stated that:

> Expenditures to improve sanitation, education and public health should, without question, be given first priority in any program to increase the long-range growth and development of the Nicaraguan economy.
>
> Without exception the mission found that in every sector of the economy, high disease rates, low standards of nutrition, and low education and training levels are the major factors inhibiting the growth of productivity. . . .
>
> The mission [members] feel more strongly on [the provision of pure water and sanitation facilities] than on any other [recommendation] presented in the report . . . pure water and sanitation facilities should take overriding priority.[12]

This advocacy had no impact on the Bank's lending policy. Of the eleven loans to Nicaragua between 1951 and 1960, not one was designated for water, sanitation, health, or education-related projects.

Table 2.2 World Bank lending before the establishment of the International Development Association (IDA), in billions of US dollars

Recipient	Gross commitments		Net lending, 1948–61[c]
	1948–61[a]	1956–61[b]	
More developed countries[d]	1.7	0.9	1.1
Colonies[e]	0.5	0.3	0.4
Less developed countries[f]	2.9	1.7	2.3
Power and transportation	2.4	1.4	2.0
Agriculture and irrigation	0.1	0.1	0.1
Total development loans	5.1	2.8	3.9

Source: World Bank, *Annual Report 1961.*

Notes
All figures exclude reconstruction loans. Sums may not total because of rounding.
a   Commitments from March 1, 1948, through April 30, 1961.
b   Commitments from July 1, 1956, through April 30, 1961.
c   Gross commitments from March 1, 1948, through April 30, 1961, less repayment of principal, cancellations, and participations and sales from portfolio to other investors.
d   Australia, Austria, Belgium, Denmark, Finland, Iceland, Israel, Italy, Japan, the Netherlands, Norway, and South Africa.
e   Algeria, Belgian Congo (Zaire), Cote d'Ivoire, Gabon, Kenya, Mauritania, Nyasaland (Malawi), Nigeria, Northern Rhodesia (Zambia), Ruanda-Urundi (Burundi), Southern Rhodesia (Zimbabwe), Tanganyika (Tanzania), and Uganda.
f   Brazil, Burma (Myanmar), Ceylon (Sri Lanka), Chile, Colombia, Costa Rica, Ecuador, El Salvador, Ethiopia, Guatemala, Haiti, Honduras, India, Iran, Iraq, Lebanon, Malaya (Malaysia), Mexico, Nicaragua, Pakistan, Panama, Paraguay, Peru, Philippines, Sudan (after independence in 1956), Thailand, Turkey, United Arab Republic (Egypt), Uruguay, and Yugoslavia.

In addition to its argument that power and transportation projects stimulated economic growth, the Bank further insisted that focusing on specific projects would enable it to directly supervise the implementation of its programs. Infrastructure projects were preferred because they enabled the Bank to assume the role of a heavy interventionist lender, and thereby assist with project preparation and implementation, dispense economic and technical advice, and make its lending contingent upon the behavior of borrowers. As Kapur *et al.* (1997: 122) note, "Visibility, verifiability, and apparent productivity were the touchstones of projecting an image of supervised, controlled, safe, quality lending." These criteria, according to the Bank, were best satisfied by large-scale, import-sensitive, long-lasting investment projects such as dams, power stations, and roads. The results of these investments could be described, photographed, and trusted; the same could not be said of Bank funds spent on intermediate goods, short-lived assets, or salaries. In this way, the Bank intended its projects to become showcases to promote "better project management" in developing countries.

The Bank's interpretation of "better project management" eventually came to mean a greater tutelary and supervisory role over the borrowers' economy. This broad role was initially seen as a reinforcement of sound lending as the Bank reasoned that underdeveloped countries needed assistance in the selection, preparation, and management of viable investment projects. But the drive to assist and advise soon went beyond the strictly ancillary requirements of Bank projects. In fact, the Bank began to pursue a more ambitious and controversial role as "tutor and influencer, using its advantageous position as a low-cost, long-term lender and its comfortable administrative budget" (ibid. 1997: 127).

Soon, as the 1950s progressed, individual projects came to be seen mostly as instruments for influencing the larger development agenda. In fact, despite the shifts in the Bank's development philosophy, this interventionist role has remained constant through a variety of programmatic initiatives, from the infrastructure projects of the 1950s, through the structural adjustment loans of the 1980s, to the recent calls for better governance.

The specific projects approach was additionally well suited to the Bank's desire to present the image of soundness. President Eugene Black explained the benefits of the approach and the supervisory role of the World Bank to a group of investors, stating that it was analogous to the US government's program of supervised credits for poor farmers during the Great Depression:

> As a result of this close combination of financial and technical assistance, practically all the loans were repaid, the fertility of millions of acres was restored, and many thousands of people were transformed from a drag on the economy into self-respecting and self-supporting producers.
> What the International Bank is trying to do is quite similar . . .

> Technical advice alone is not sufficient . . . nor is financial assistance. What is needed is a combination of the two.[13]

Recounting the many precautions taken by the US government's program to ensure repayment, Black tried to convince his listeners that, with tough standards and hands-on lending, one could reconcile good business with doing good. The rationale for the Bank's project focus, tutelary role, and interventionist lending style was developed in successive speeches and annual reports. The *Fifth Annual Report* explained that the specific project approach was a "safeguard" designed simply to assure that Bank loans would be used for productive purposes:

> If the Bank were to make loans for unspecified purposes or vague development programs which have not been worked out in terms of the specific projects by which the objectives of the program are to be achieved, there would be a danger that the Bank's resources would be used either for projects which are economically or technically unsound or are of a low priority nature, or for economically unjustified consumer goods imports.[14]

The Bank's concentration on infrastructure-related financing was influenced by schools of thought, such as modernization (discussed in Chapter 1), that regarded external capital investment as a locomotive for growth, as well as the groundwork for private sector capital investment, which could not be attracted to areas that lacked basic infrastructure (Mason and Asher 1973). Thus, the Bank was swimming in the intellectual currents of the 1950s, with its lending programs designed to bring about overall economic growth in national economies, rather than to address social needs or poverty. Its logic was premised on the notion that economic growth was the best remedy for poverty in developing countries, and that efforts to cut this process short by funding socially oriented projects would be counter-productive. However, the changing political climate of the 1960s pushed the Bank toward the very social concerns it was initially reluctant to address.

Multilateral development organizations and social lending: the politics of development assistance

Throughout the 1950s, while concentrating on financing large-scale infrastructure projects, such as dams and highways, the Bank insulated itself from outside criticism and political pressure to alter its funding focus. However, its narrow, self-interested approach to development assistance was increasingly questioned during the late 1950s and early 1960s. It was argued in many circles that the Bank's lending bypassed the poor and ignored social justice in the developing world (United Nations 1951; Viner 1953). Meanwhile, Third World nations were becoming more vocal in international forums such as the United Nations, and their numbers within those organizations increased as a consequence of decolonization. The Bank could no longer turn a deaf ear to their calls to broaden its lending program.

By the early 1960s, the Bank found itself in a rapidly changing international environment. Third World nations had campaigned vigorously during the mid to late 1950s for a fund, under the auspices of the United Nations, to provide for economic and social development. By the late 1950s, both the Soviet Union and the United States had come to believe that their own survival depended on "winning over" Third World nations. One implication of the Cold War, according to then Secretary of State John Foster Dulles,[15] was the need "to make political loans and soft loans on a long term basis" (cited in LaFeber 1993: 177). Against the objections of conservatives, Dulles argued for "soft" lending (i.e. lending with fewer conditions), warning that "it might be good banking to put South America through the wringer, but it will come out red" (cited ibid.). Entangled in the politics of aid during the Cold War, the World Bank had to adjust its lending program accordingly. Furthermore, the Bank's orthodox model of economic development was itself under increasing interrogation. Therefore, at the beginning of the 1960s, the Bank was struggling to hold on to its image as a market-based, financially conservative institution, while simultaneously fielding the social, political, and intellectual demands of the 1960s.

Third World concerns

As noted above, Third World countries used their increasing numbers within the United Nations to articulate their needs and concerns.[16] In a March 1949 report, the United Nations Sub-commission on Economic Development noted the following:

> on a realistic assessment it cannot be assumed that the Bank could, in the foreseeable future, be able to make a significant contribution to the massive investments required for economic development involved over a long period. Moreover, even if the finance available through the Bank could be increased beyond expectations, the Sub-Commission is of the opinion that the terms on which it would be available under the policy established by the Bank limit the effectiveness of this financing to under-developed countries. There are fields and types of investment required for economic development which can neither satisfy the pre-conditions required by the Bank, nor carry the interest charges involved, nor be liquidated within the period required . . .
> It is for these reasons that the Sub-commission extended its consideration of international finance beyond the activities of the Bank,

and discussed the possibilities of opening new sources of international finance under United Nations auspices.[17]

The chair of the commission, Mr. V.K.R.V. Rao, attached an appendix to the report suggesting that the new international organization be called the United Nations Economic Development Administration (UNEDA). This organization was to be funded by contributions from member governments, in their own currencies, and would perform a variety of development-related functions. Financial assistance from UNEDA would "normally take the form of loans and not grants, though the terms of repayment would be liberal, and the interest charged may be only nominal" (Mason and Asher 1973: 382).

However, the American representative to the Sub-commission, Emilio G. Gollado, who had previously served as the first US executive director of the Bank, was "unable to concur" with the Sub-commission's report. The American government's view, according to Gollado, was that nations should

"look primarily to American private enterprise"; when development assistance could not be obtained through private sources, they "should rely fundamentally on the International Bank for Reconstruction and Development for financing or collaborating in financing closely circumscribed types of projects basic to development and not readily susceptible to implementation by purely private financing."[18]

Furthermore, the World Bank also was not enthusiastic about Rao's report, because the proposed functions of UNEDA either came within the Bank's terms of reference or did not need to be performed, "since loans made by the Bank are on terms which are not designed to make any substantial profit."[19] The Bank also maintained that "any greater liberality in terms of UNEDA loans would amount simply to disguised inter-governmental grants."[20] Bank president Eugene Black himself voiced similar objections to such loans:

loans of this kind are ... part loan and part grant. They ... are not always apt to be regarded as serious debt obligations. Like other fuzzy transactions, they therefore tend to impair the integrity of all international credit operations.

(cited in Kapur et al. 1997: 136)

However, despite the United States' and the Bank's unenthusiastic response to UNEDA, Third World nations remained steadfast in their pursuit of a development fund. The debate resurfaced as a proposal for an International Development Authority, again within the United Nations, as a Special United Nations Fund for Economic Development (SUNFED) (Hadwen and Kaufmann 1960; Weaver 1965). Again, after opposing the idea, stalling, and trying to deter the Third World campaign, the United States eventually tried to co-opt the idea by giving it tacit support. The World Bank gradually followed suit and shifted its own antagonistic position toward the development fund. A consensus emerged, within both the Bank and the US government, on the need to resist LDCs' pressures for a soft loan agency under the control of the United Nations, where the developing nations had greater voice (Baldwin 1966; Payer 1982). Black changed his mind on "fuzzy loans" and agreed to the creation of an International Development Association (IDA) in 1959. After much debate, discussion, and dissent on the function, bureaucratic location, and structure of this development fund, the IDA was finally attached to the World Bank. Thus, the IDA, the soft loan affiliate of the World Bank, formally came into existence in 1960.[21]

The formation of the IDA increased the number of nations with which the Bank dealt, as well as development projects. The IDA was formed at a time when the Bank's international role itself was uncertain. Western Europe and Australia were becoming too creditworthy to continue borrowing from the Bank; Japan, while still a borrower at this time, was emerging as a strong economic power in its own right. Thus, the institutionalization of the IDA in 1960 ensured that the Bank could continue playing an important role in international development during a period of political flux, when its own *raison d'être* was being called into question.

Intended as a substitute for SUNFED, the general purpose of the IDA was to offer softer, less conditioned development assistance to Third World nations. The proposal and charter of the IDA, however, were silent on the issue of allocation; it was left up to the Bank to decide for whom, and for what, IDA money would be used. As a result, the Bank was acutely aware that much of the political expectations from SUNFED were transferred onto the IDA. In November 1960, a staff economist visiting the Middle East on a field mission noted that

> There continue to be misapprehensions about the IDA. . . . [When we] visit a town or village, the mayor and town council produce a list of projects. . . . After striving for fifteen years to achieve Wall Street respectability, the Bank watched as the IDA suddenly materialized and conjured up the 1940s' augury that the Bank would grow up to be a soup kitchen.
>
> (cited in Kapur *et al.* 1997: 99)

An internal identity crisis had thus been introduced into the Bank. In spite of the IDA's proposed role and external demands, the Bank wanted to maintain its conservative image as a strong bank, and not a social welfare organization or grant-giving agency. In the October 1959 annual meeting, Black pledged that the "IDA will not be a soft lender."[22] This pledge, however, seemed contradictory and out of place and was difficult to meet, given the reality surrounding the formation of the IDA. The Bank's management eventually decided that it would not be in their best political interest to totally exclude social lending; instead, they would simply give a diplomatic hint that they would not fund it. The management thought it prudent to summarize its position to the board of governors as follows:

> IDA's financing would be largely concerned with directly productive projects of the type normally financed by the Bank, but . . . social projects would not be excluded. We would prefer to avoid any reference to health and education projects.
>
> (ibid.: 156)

However, Bank governors from the Third World had assumed that the IDA would lend to projects not covered by general Bank lending. For instance, the governor from Ecuador emphasized that the IDA "should have in mind the social conditions of the people," and lend for "housing, school buildings, education of the masses, public health and sanitation [which] have a major and fundamental effect on economic development."[23] Black sought to quell these expectations in his own closing remarks, but his efforts were undermined by US delegate Douglas Dillon, who welcomed the IDA's role as an opportunity to broaden the lending facilities available to the Bank so that "it may play its part more effectively in the historic struggle of man against poverty and disease"; furthermore, Dillon pointed out that "high technical standards" of IDA projects would not rule out "financing pilot projects in some fields of social overhead."[24] The US executive director at the Bank, T. Grayton Upton, also added that "pilot projects in the field of social development might be appropriate for the IDA," and though "the United States would emphasize productive projects of an economic character" it "recognizes the strong interest on the part of several countries in the financing of the so-called "social projects."[25] Thus, as a result of both the demands of the Third World and the geo-political concerns of the United States, the reluctant Bank management was increasingly pushed in the direction of social lending.

The Bank's shift is illustrated by the IDA loan to Jordan. The loan committee had initially deemed Jordan to be uncreditworthy for a water supply loan in February 1958. However, in light of a crisis in the Middle East five months later, when a pro-Nasser army coup occurred in Iraq, the World Bank re-evaluated its position on lending to Jordan in October 1960 as part of an attempt to prevent a similar coup in Jordan, and in order to protect King Hussein, an American ally. A few weeks before the IDA even opened its doors for business, the Loan Committee had already considered a credit application for water supply in Amman. Despite the fact that it had not previously ventured into the area of funding water supply, the Bank nevertheless funded the loan (Kapur *et al.* 1997).

The debates over Jordan's loan application, and those from other Cold War sites in Latin America, ignited a more general discussion within the Bank on the sectoral allocation of its resources. The rapid acceptance of Jordan's request had set a new precedent, which caused the Bank to rethink its once antagonistic position on funding water supply loans. In October 1960, the staff began to discuss future policy on water supply, and by November 7 the Technical Operations Department (TOD) had submitted a draft policy statement to the Staff Loan Committee (SLC) foreseeing large potential benefits: "Few projects ... incorporate as great a potential for directly benefiting the vast majority of the people ... as does water supply improvement."[26] Robert Sadove, director of special projects, observed that water was an important industrial output and thus, "not essentially different from electric power."[27] Thus, it was through the IDA that the Bank ventured into sectors that it had traditionally shunned.

The Cold War and geo-political considerations

The formation of the IDA not only enabled the Bank to venture into social lending; it also enabled donor governments, particularly the United States, to insert their own political objectives into the Bank's agenda. As voting at the Bank was based on capital subscriptions, the United States, as the largest subscriber, was able to exercise a great degree of control over the use of its contributions (Gwin 1994). Development assistance thus became an instrument of Cold War politics as the United States tried to nurture its spheres of influence in the Third World (McNeill 1981; Bhagwati 1985; J.M. Cohen et al. 1985; Conteh-Morgan 1990). Black was apprehensive about this trend and objected to the growing linkage between development aid and Cold War objectives:

> Diplomatists and strategists [who] offer economic aid in exchange for a military alliance or a diplomatic concession ... are certainly not serving the interests of orderly economic development; in fact they may well be abetting and perpetuating conditions which in the short [run] will render their military alliances and diplomatic concessions quite hollow victories.
>
> (Black 1960: 268–9)

However, his arguments met little success in the Cold War atmosphere. Dulles had stated earlier, in 1956, that "East and West are in a contest in the field of development of under-developed countries ... Defeat ... could be as disastrous as defeat in the arms race" (cited in Clifton 1992: 776). The foreign policy establishment in the United States became especially concerned about the convergence of nationalism and socialism in many parts of the Third World. National liberation movements in Africa, Asia, and the Caribbean, as well as Leftist political movements in Latin America, were attracted to schools of thought that emphasized the structural imbalances and dependency between First and Third World nations as a cause of underdevelopment. Some of these movements displayed an open hostility to Western, pro-capitalist models of development, and were captivated by the Soviet experience of socialist transition, as well as the model of Chinese participatory, egalitarian, rural development (Fanon 1963; Sigmund 1972). An alarm was set off in the West by Fidel Castro's victory in Cuba and the emergence of communism in America's "backyard." The United States tried various covert measures to depose Castro in Cuba and Rafael Trujillo in the Dominican Republic, while the United Kingdom rigged elections in British Guyana to block the Marxist candidate, Cheddi Jagan. Cold War geography brought about a shift in American foreign policy from the localized containment of communism to a generalized competition for political allegiance, of which efforts to forestall communism through economic and social development assistance became a part.

Western policy-makers – Americans in particular – increasingly realized that the threat of socialism and communism could not be nullified by military measures alone. The large-scale socioeconomic inequalities that characterized the Third World would have to be addressed in some fashion in order to make socialism seem less attractive. Thus, the World Bank was part of a political tide which increased not only the volume and urgency of development aid, but also the sectoral distribution of that aid. As a result, poverty in developing countries came into sharper focus and the Bank's traditional arguments for "patience" and economic growth were softened. The Bank began to accept the need to make some socially oriented concessions in lending – which policy-makers understood to mean welfare-related investments in housing, water supplies, health services, and education – in order to prevent socialist revolutions.

Bank vice-president Robert Garner, who was responsible for the Bank's fiscally conservative philosophy during the 1950s, reflected on this new scenario in his farewell address:

> Feudal society, with its wealth and power in the hands of a few . . . must disappear if there is to be economic progress . . . So I put high on the list of public policy, positive efforts to see the benefits of growth spread more widely.[28]

Speaking to the Senate Foreign Relations Committee in the aftermath of the Cuban revolution, American Undersecretary of State Dillon[29] also pointed to the distributive failures of previous development efforts: "While there has been a steady rise in national incomes throughout Latin America, millions of under-privileged have not benefited" (cited in M.S. Eisenhower 1965: 249). It was in this context that the IDA made several loans to Latin American countries in the early 1960s. Honduras, Chile, Nicaragua, Columbia, Costa Rica, and Paraguay all received support from the IDA during its first two years of its operation. American interest in Latin America increased following Castro's sweeping nationalizations and expropriations, as well as Cuba's trade pact with the USSR. Returning from his trip to South America, US president Eisenhower stated that:

I am determined to begin . . . historic measures designed to bring about social reforms for the benefit of all the people of Latin America.

Constantly before us was the question of what could be done about revolutionary ferment in the world . . . We needed new policies that would reach the seat of the trouble, the seething unrest of the people. . . . One suggestion was . . . to raise the pay of the teachers and start hundreds of vocational schools . . . [We] had to disabuse ourselves of some old ideas . . . to keep the Free World from going up in flames.

(cited in D.D. Eisenhower 1965: 530–7)

In 1961, Kennedy identified Latin America as, next to Berlin, one of the most critical areas of concern for the United States. Fear mounted that Castro might take over the whole hemisphere (Packenham 1973; McCormick 1989). In March of 1961, Kennedy demanded action to avert chaos in Bolivia. His staff

ignore[d] proposals by the International Monetary Fund that Bolivia needed a good dose of an anti-inflationary austerity, and instead offer[ed] immediate economic assistance.

(Goodwin 1988: 147)

A week later, J.F.K. announced the Alliance for Progress with Latin America, a ten-year program for cooperation and development stressing social reform, with large-scale aid to countries that cooperated with the West. The Alliance for Progress with Latin America was aimed at luring nations away from Castroism. In his address to the Latin American Diplomatic Corps in March 1961 announcing the Alliance for Progress, Kennedy said that in addition to stimulating economic growth, development aid should

combat illiteracy, improve the productivity and use of their land, wipe out disease, attack archaic tax and land tenure structures, provide educational opportunities, and offer a broad range of projects designed to make benefits of increasing abundance available to all.

(cited in Sorensen 1988: 351–2)

The Alliance thus proposed specific policy reforms in agriculture, health, housing, and education and even stressed equity issues.

Accordingly, as the Cold War escalated in the 1960s, World Bank lending was influenced increasingly by anti-communist political priorities that pushed the institution toward social lending. Given the West's perception that socioeconomic underdevelopment ignited socialist sympathies in Third World countries, addressing underdevelopment became a matter of American national security, and appropriating socialist development agendas became a method of preventing socialist revolution. There was an urgent need for, and great pressure on, institutions such as the World Bank to be involved in directly addressing underdevelopment in Third World countries. Thus, the development mission of the Bank was now charged with the Cold War's sense of political urgency.

The World Bank and development policy debates

The previous sections discussed Third World nations' campaign for an international development fund and the politics of the Cold War; in addition to these factors, the intellectual interrogation of the traditional/orthodox view emphasizing economic growth influenced the Bank to reconsider its own position on development and social lending. During the 1960s, it was widely argued in policy and scholarly circles that the traditional model of economic development paid very little attention to issues of social justice and structural imbalance between the developed and developing worlds (Ward 1962; Wilber 1979; Bloomstrom and Hettne 1984). Furthermore, in spite of impressive economic growth rates in some Third World countries, disparity between rich and poor increased and conditions of poverty worsened. In a speech to the British Overseas Development Institute in 1965 (five years into the first United Nations Development Decade), Barbara Ward[30] summed up the frustrations of individuals/agencies dealing with underdevelopment in Third World countries:

> Such were the aims five years ago ... Let us begin by trying to see where we are now, half way through the Decade. In some ways, it has not gone too badly ... Yet at the end of five years, the gap between rich nations and poorer nations is greater still, not because poorer nations have necessarily grown poorer, but because the rich have got richer by so much more.[31]

Thus, for many, simply stimulating economic growth was insufficient. Additional efforts were required to reduce poverty, unemployment and inequality. Seers, introduced earlier, and a prominent representative of this view, states it plainly:

> The questions to ask about a country's development are therefore: What has been happening to poverty? What has been happening to unemployment? What has been happening to inequality? If all three of these have become less severe, then beyond doubt there has been a period of development for the country concerned. If one or two of these problems have been growing worse, and especially if all three have, it would be strange to call the result "development," even if *per capita* income has soared.
>
> (Seers 1969: 3)

The notion that development should entail more than the maximization of economic growth began to gain ascendancy in the literature, and a number of reports by NGOs emphasized the need to re-evaluate development strategy itself. In 1965, the Dag Hammarskjold Foundation published a report titled *What Now: Another Development?* arguing that the satisfaction of basic needs should be at the core of the development process and that development should be need oriented, endogenous, self-reliant, ecologically sound, and contribute toward the transformation of social structures. In 1961, Hans Singer spoke of "a shift in our whole thinking about development . . . from physical to human capital" (Singer 1964: 46). American Economic Association (AEA) president Theodore Shultz's embrace of the term "human capital" further energized the critique of the orthodox view of development. In his presidential address to the annual meeting of the AEA in December 1960, Shultz criticized the World Bank's myopic view of development:

> [The World Bank was] responsible for the one-sided effort to transfer physical capital alone to the developing countries in spite of the fact that . . . knowledge and skills [are] the most valuable resource that we could make available to them . . . [By its] export doctrines, the Bank contributed to the neglect of human capital.
>
> (Shultz 1961: 11)

The critique of the traditional models of economic development and the new concept of investing in people were quickly adopted by agencies such as the United Nations Educational, Scientific and Cultural Organization (UNESCO), the International Labor Organization (ILO), etc. As Benjamin Higgins,[32] who was then working for UNESCO, recalls:

> When the "residual factor" burst on the scene . . . UNESCO was quick to say, "the residual factor, of course, is education." ILO was equally quick to add, "true, but a major component of education for development is manpower training." FAO stressed the importance of training farmers. WHO was a bit slow in pointing out that the "residual factor" might include improvements in nutrition and health as well.
>
> (Higgins 1989: 97)

In 1963, Jan Tinbergen persuaded the Dutch government to finance the creation of the United Nations Institute for Social Development (UNISD). The World Bank, fearing isolation from this intellectual and policy climate, gradually tried to include social lending in its programmatic initiatives in order to stay relevant. However, the Bank always filtered contemporary debate on development through its own lenses, in characteristic fashion, in order to retain its hegemonic status in the realm of ideas as well as politics (discussed in Chapter 1).

It is apparent that it was against the background of the Cold War, the formation of the IDA, and changing intellectual currents in development studies that Robert McNamara began to shift the focus of the Bank's lending program in the late 1960s.

The McNamara years: the World Bank and poverty alleviation

Robert McNamara was, and remains, a complex personality. His energetic and controversial leadership of the World Bank from 1968 to 1981 impressed its stamp on the institution, transforming its culture and operations. While the Bank's evolving concern with poverty-related issues ought to be understood in the context of the geo-political scene described above, it is also important to acknowledge how McNamara's leadership also transformed the World Bank from within. While structural circumstances pertaining to the Cold War did govern the unfolding of the Bank's poverty-related programs, the role of actors within and outside the Bank in resisting, modifying, or conforming to these structures should not be overlooked. While history is not predetermined, it is important to note that neither are the actions in history random, haphazard, or purely coincidental. These actions, which are influenced by the catalogue of past actions, which are in turn created by both structures and agents, can then lead to a variety of outcomes. In his pamphlet *The Eighteenth Brumaire of Louis Bonaparte*, Marx captured this nuance:

> Men make their own history, but they do not make it just as they please; they do not make it under circumstances chosen by themselves but under circumstances directly encountered, given and transmitted from the past.
>
> (cited in Beauregard 1984: 64)

Echoing Marx, Beauregard also observes that

> History and structure are not "given" or pre-ordained by some mystical force or spirit. Rather, people make history and structure, even as history and structure contextualize their actions.
>
> (ibid.: 62)

As noted above, the geo-political climate and the general international demand for appropriate programs to address the development gap between north and south undoubtedly influenced the World Bank's outlook toward socially oriented lending. However, the way in which actors within the Bank responded to these structured circumstances also contributed to the Bank's reorientation. It is in terms of this exchange between structures and actors, between circumstances and individuals, that Robert McNamara's presidency of the World Bank ought to be understood.

McNamara assumed leadership of the World Bank in 1968, immediately after serving seven years as the US Secretary of Defense.[33] According to World Bank historian Jochen Kraske, McNamara had a moralist approach to development and he "believed in the ability of men and institutions to solve problems."[34] George and Sabelli (1994) add that "McNamara unquestionably brought a capacity for hard work, zeal, self-righteousness, discipline and commitment to helping the poor" to the presidency. However, his tenure was punctuated with controversy because of his role in the Vietnam War. In addition to his awareness of the power of aid as an instrument of foreign policy and diplomacy, some commentators speculate that McNamara may have become a strong advocate for socially oriented lending because of his tragic sense of failure and the need for redemption from the national trauma of the war (ibid.).

Nevertheless, McNamara, as the youngest appointee to the position at the age of fifty-one, brought a sense of vision and persuasion to the job. At the end of his first week at the Bank, McNamara called the President's Council (the most senior Bank officials) to ask them these questions: Why was the lending for this fiscal year going to be below a billion dollars? Why are so many needy countries neglected by the Bank? The reasons given were numerous: Indonesia had only recently returned to the Bank's fold; Nasser's Egypt was unpopular with the US Congress; and most countries in Africa were considered too backward for appropriate projects that met World Bank standards, and so on. McNamara listened with some impatience as the gloomy tenor of the meeting was deepened by a sudden, premature darkening of the sky outside (Shapley 1992). He ended the meeting abruptly, stating, "I am going to ask you all to give me very shortly a list of all the projects that you wish you could see the Bank carry out if there were no financial constraints" (cited in Clark 1981: 168).

The Council filed out of the room astonished by McNamara's proposal, which seemed to defy, to the letter, the fiscally conservative principles that the Bank had nurtured over the years. The gravity of the situation was soon compounded by the discovery that the darkness outside was caused not by a passing thundercloud, but by smoke, as large numbers of the Washington poor erupted in outrage over the assassination of Martin Luther King, Jr.

For the next few months, the Bank staff worked tirelessly to come up with ideas that would meet McNamara's challenge to expand and reorient World Bank lending. In time, two major changes were initiated under the McNamara presidency: (i) an expansion in the flow of financial resources from the Bank to countries in the developing world; and (ii) a re-orientation in the types of projects financed by the Bank. It must be noted here, however, that, while the Bank began to venture into new fields, it still maintained an overwhelming interest in basic infrastructure projects initiated prior to McNamara's presidency (Table 2.3).

The changes introduced by McNamara affected four specific areas of the Bank's operations:

1 Lending increased. In 1968, when McNamara assumed the presidency of the Bank, sixty-two new projects were approved by the Bank and IDA. By 1981, some 266 projects had been approved.
2 There was a shift away from infrastructure projects toward anti-poverty programs. Lending increased for rural poverty programs, low-income urban housing, slum rehabilitation, small-scale industry, primary school education, as well as health and nutrition programs.
3 The percentage of staff from developing countries increased.
4 There was a reorganization of the Bank itself. Previously, the Bank had a bipolar structure with client-oriented geographic departments, and technical operations divisions containing country operations. In order to watch over general standards, divide skills across regions, and accommodate departments that were hard to regionalize, certain departments were operated on a sector basis under a separate vice presidency headed by Warren Baum. The urban division was established and operated under this column.

(Ayres 1983: 4–7)

While he endorsed the Bank's insistence on the overriding need for economic growth, McNamara argued that economic growth and poverty reduction could no longer be considered synonyms. In his first speech at the 1968 Annual General Meeting, he noted that since 1960, in the developing world,

the average annual growth thus far has been 4.8 per cent ... And yet ... you know that these cheerful statistics are cosmetics which conceal a far less cheerful picture ... Much of the growth is concentrated in the industrial areas, while the peasant remains stuck in his immemorial poverty, living on the bare margin of subsistence.

(McNamara 1981: 3–5)

In this statement, McNamara challenged a key article of the Bank's creed, which held that rising national income in LDCs would benefit their poorer citizens.

McNamara's advocacy for the poor is best understood when contextualized within the historical period of his presidency, which offers three reasons why McNamara steered the Bank toward social lending: national security concerns, the critique of the aid establishment, and issues/events that influenced him personally. These are discussed below.

Table 2.3 Distribution of Bank group lending to LDCs, by sector, 1960–71 (gross commitments expressed in millions of US dollars)

	Fiscal years									
	1960		1965		1970		1961–70		1971	
Sector	Amount	Percent	Amount	Percent	Amount	Percent	Amount	Percent	Amount	Percent
Economic										
Agriculture	46	9	167	15	423	18	1,689	15	428	18
Industry	118	22	119	10	403	18	1,842	16	248	11
Transportation	200	38	422	37	664	29	3,364	29	625	27
Communications	–	–	33	3	85	4	362	3	196	8
Power	165	31	335	30	529	23	3,011	26	480	20
Tourism	–	–	–	–	1	–	18	–	10	0
Subtotal	529	100	1,076	95	2,105	92	10,286	89	1,987	84
Social										
Education	–	–	30	3	80	3	326	3	100	4
Health	–	–	–	–	–	–	–	–	–	–
Water supply	–	–	27	2	33	2	250	2	154	7
Family planning	–	–	–	–	2	–	2	–	8	0
Subtotal	–	–	57	5	115	5	578	5	262	11
Maintenance imports	–	–	–	–	75	3	655	6	25	1
Special and other	–	–	–	–	3	–	5	–	85	4
Total	529	100	1,133	100	2,298	100	11,524	100	2,359	100

Source: IBRD, Economic Program Department.

National security concerns

McNamara believed that national security and world poverty were closely related. In a speech before the American Society of Newspaper Editors in Montreal, he argued that "Security is not military hardware . . . without development there can be no security."[35] McNamara, like other US officials at the time, was obsessed by the communist threat to the stability and power of the United States. He felt that revolution in any place on earth, regardless of how poverty-stricken and obscure the country, imperiled the "free" world. In a testimony before the US Congress in 1969, McNamara emphasized that:

> The death of Ernesto Che Guevara in Bolivia in the fall of 1967 dealt a severe blow to the hopes of the Castroite revolutionaries. But counterinsurgency alone is an inadequate response to this problem. Removal of the causes of human suffering and deprivation is essential if stable political institutions are to flourish free of the threat of violent revolution.[36]

In his book, *The Essence of Security*, McNamara (1968: 109–10) explains that "a nation can reach a point at which it does not buy more security for itself simply by buying more military hardware and we are at that point." The threat to the United States and its allies comes from those "traditionally listless areas of the world [which have become] seething cauldrons of change. In dealing with them, sophisticated weapons and more defense dollars will get you nowhere" (ibid.: 115).

McNamara characteristically hammered home this point with figures, citing the "164 internationally significant outbreaks of violence" in the previous eight years (1958–66), "each of them specifically designed as a serious challenge to the authority or the very existence of the government in question" (cited in Shapley 1992: 429).[37] These outbreaks of violence were not classic cross-border wars, but large-scale, internal, civil insurgencies. The governments under threat in each case were allies of the United States, which worried McNamara.

At the US Congressional hearings on the 1969–73 defense budget, McNamara testified that:

> We could find ourselves literally isolated, a "fortress America" still relatively prosperous but surrounded by a sea of struggling, envious, and unfriendly nations – a situation hardly likely to strengthen our own state of peace and security . . . We must create conditions for economic and social progress in the less developed areas of the world.
>
> There is a direct and constant relationship between the incidence of violence and the economic status of the countries afflicted . . . since 1958, 87 per cent of the very poor nations and 48 per cent of the middle-income

nations suffered serious violence ... there is a relationship between violence and economic backwardness and the trend of such violence is up, not down.

(McNamara 1968: 7)

McNamara assumed that development could be achieved through outside intervention and saw the Bank's focus on social lending as a means to abate the socialist threat stirring in the Third World. Thus, his advocacy for World Bank poverty-directed lending paralleled the thinking of foreign policy-makers like Dulles, mentioned earlier, who argued that socioeconomic problems in Third World countries ought to be considered a national security concern of the United States.

Critique of the aid establishment

There was a great deal of frustration within the aid community regarding the increasing disparity between the rich and the poor despite impressive economic growth rates. A number of forums in the late 1960s questioned the conventional wisdom of aid programs and their impact on the poor. During the early 1960s, McNamara's predecessor at the Bank, George Woods, had begun to argue in some of his public addresses that the Bank ought to concern itself with the formation of human capital in addition to investment in physical infrastructure alone. In October 1967, Woods suggested the formation of a commission to study the results of twenty years of development assistance, clarify the errors, and propose policies that would improve results in the future. On arrival, McNamara heeded Wood's call and invited Lester B. Pearson, the former prime minister of Canada, to undertake such a study. The Pearson Commission was mostly concerned with the operation and consequences of foreign aid, and how it could more effectively contribute toward the complex business of economic development (Byres 1972). Setting the stage for discussions on aid programs in the 1970s, the Pearson Commission's report acknowledged that, despite the advances made by developing countries during the 1950s and 1960s, they still faced numerous challenges in meeting their development requirements. The completed report, titled *Partners in Development: Report of the Commission on International Development* and submitted to McNamara on September 15, 1969, addressed several issues pertaining to making development assistance more effective (see Table 2.4 for a summary).

The recommendations of the Pearson Commission received the immediate and intensive attention of the World Bank management. McNamara stated in his first address to the board of governors in 1968 that the Pearson Commission report would enable the Bank to develop effective strategies to

Table 2.4 Summary of the Pearson Report

Political Reform	The Commission noted that one of the major problems was the concentration of the benefits of economic development in the hands of a few. The report recommended that policies designed to redistribute income be given the same priority as those designed to accelerate growth. Land and administrative reforms were also recommended in order to address the concerns of the poor
Population growth rates	These were identified as "the major cause of the large discrepancy between rates of economic improvement in rich and poor countries" (p. 55). The report contended that high population growth rates create budgetary strains by increasing expenditure on education, health, housing, water supply, etc. This diverts scarce government resources to a dependent population that would otherwise have been used to raise standards and increase capital formation. While the report acknowledges the politically sensitive nature of recommending curtailing growth in Third World societies, it stresses that "there can be no serious social and economic planning unless the ominous implications of uncontrolled population growth are understood and acted upon" (p. 58)
Unemployment and urbanization	These were identified as major obstacles to development. The report recommended strengthening agricultural sectors and adopting a cautious mechanization program in agriculture in order not to displace agricultural workers. Urbanization was briefly mentioned in the report and recommendations were made for the promotion of small and intermediate regional centers
Agriculture	The Green Revolution is identified as an important breakthrough in food grain production. Areas untouched by the Green Revolution faced the difficult challenge of stimulating technological change in the country side, according to the commission. A program of structural change in land ownership was recommended by the report
Industrial policies	The Commission mentioned that many developing countries tend to favor industrialization to the detriment of the agricultural sector. Import substitution industrialization was seen as a problem because many Third World countries found themselves with a highly distorted price structure, making them non-competitive in export markets
Private sector development and research and development	The lack of a dynamic private sector was identified as an impediment to development. Nationalistic concerns over the control of the economy created an uncertain atmosphere for private sector investment. The report concluded that the public sector should facilitate the development of private business and that a vigorous private sector is an important element in stimulating economic growth. The report also called for increased resources for research and development and the restructuring of educational programs to meet the developmental needs of a given country

External constraints	The Commission's report acknowledged that Third World countries' development policies are pursued within an international context over which they have very little or no control. The major external constraints identified by the report are the availability of foreign currency, the debt problem and unequal trading relationships with First World countries. Recommendations were made for an improvement in terms of trade between First and Third World countries and the removal of barriers against the export of goods from developing countries. Furthermore, the volume of official aid from the developed world should increase to 0.7 percent of the gross national product. This increase should be combined with better partnerships, clearer purpose, and greater cohesion in administration

Source: Pearson (1969: various pages).

address the problems in developing societies, "not just for the next decade, but for a whole generation that will carry us to the end of this century."[38]

In September 1970, McNamara endorsed the general principles of the report, specifically mentioning that:

> Economic development alone would not be enough to accomplish development objectives. In addition to economic growth, problems relating to population growth, rural under-development, . . . international development assistance was crucial for social stability in third world countries.
>
> (McNamara 1981: 114–16)

These examples illustrate that McNamara's leadership propelled an initially reluctant World Bank into the vanguard of social lending.

Personal and intellectual influences

Influences on McNamara may be traced to several personal and intellectual sources. His wife, Margaret, was actively involved in Reading is Fundamental, an organization that aimed to address the problems of illiteracy in the developing world. McNamara's biographer, Deborah Shapley (1992), notes that Margaret McNamara lobbied her husband to take on specific social issues at the Bank. McNamara's assistant, William Clark (mentioned earlier), a Fabian socialist, introduced McNamara to acquaintances who attacked the aid establishment and argued that current programs benefited only the elite in Third World societies. For example, Clark's close friend, Barbara Ward (also mentioned earlier), was one of the most vocal critics of the aid establishment in the United States during the 1960s. According to Clark (1981), McNamara soon began to send drafts of his speeches to Ward, and her ideas came to constitute an intellectual framework for McNamara's anti-poverty programs in developing countries. McNamara himself has said, "She influenced me more than anyone in my life" (Shapley 1992: 507).

In 1970, Ward organized a conference at Columbia University which brought together various critics of the aid establishment, such as United Nations Development Programme (UNDP) director Paul Hoffman, Johannes Wittveen, director of the International Monetary Fund, and Maurice Strong, the head of Canada's aid program. She also invited Robert McNamara to attend. In his address to the conference, McNamara predicted that the faultline that divides the world would shift from an east–west axis to a north–south axis if intervention in Third World poverty were delayed any further. He also argued that the criteria for evaluating development should entail more than gross measures of economic growth, a theme that became very important during his tenure at the Bank. At the conference, McNamara vigorously engaged participants on how World Bank lending might target the poor directly.

Also attending the conference was Mahbub ul Haq, a Pakistani economist who had served on his country's planning commission for thirteen years. Ul Haq was an ardent critic of the Western aid establishment who did not hesitate to voice his dissent.[39] Attracted to ul Haq's defiant critique, McNamara asked him to submit a memo outlining how the Bank's lending could focus directly on the poor. Ul Haq initially thought it was a trick: "I felt that probably he wanted me to commit myself in writing and so demonstrate just how shoddy some of the arguments were" (cited in Shapley 1992: 508). However, McNamara proved his interest in the critique by subsequently bringing ul Haq and others into the Bank as part of the president's inner core. This core developed specific policies to meet McNamara's broad goals of poverty alleviation in the developing world.

Toward social lending

The first term of McNamara's presidency, from 1968 to 1973, may be characterized as a "time of intellectual and operational gestation" (Kapur *et al.* 1997: 209) during which the Bank, through studies, consultation, and experimentation, sought to define poverty-oriented lending and, above all, design suitable programs with a social bent. By the time of his annual address to the Bank's board of governors in 1973, McNamara had developed an agenda and set specific goals for his poverty alleviation program. The meeting was held in Nairobi, Kenya, and marked the first time a World Bank annual meeting was held on the African continent. In his address, McNamara outlined an ambitious program for addressing poverty and improving the productivity of the rural poor. The "Nairobi address," as it came to be known, was regarded as a watershed by the development community and stands as one of McNamara's most influential policy statements.

At Nairobi, McNamara differentiated between relative and absolute poverty:

Relative poverty means simply that some countries are less affluent than other countries, or that citizens of a given country have less personal abundance than their neighbors. That has always been the case, and granted the realities of differences between regions and between individuals, will continue to be the case. But absolute poverty is a condition of life so degraded by disease, illiteracy, malnutrition, and squalor as to deny its victims basic human necessities, . . . a condition of life so limited as to prevent the realization of the potential with which one is born; a condition of life so common as to be the lot of some 40 per cent of the peoples of developing countries.

(McNamara 1981: 237–8)

McNamara's plea to help the "absolute poor" was well received in development circles. Clark (1981: 177) observes that the phrase in fact served another end: it was McNamara's answer to those who argued that developed nations could not afford the increased amounts of funding he requested because of their own domestic poverty concerns. McNamara tried to show that, although the poor in developed countries needed attention, their poverty was "relative" compared to the "absolute poverty" found in the Third World.

The basic problem of poverty, McNamara (1981: 242) surmised, is that "growth is not equitably reaching the poor and the poor are not significantly contributing to growth." The problem of poverty revolved around the "low productivity" of the rural poor, according to McNamara. While he acknowledged that there were no clear answers to this issue, he was determined that the World Bank should make a beginning. Essential to any comprehensive strategy to increase the productivity of smallholder agriculture, he said, were the following:

acceleration in the rate of land and tenure reform; better access to credit; assured availability of water; expanded extension facilities backed by intensified agricultural research; greater access to public services and new forms of rural institutions and organizations that will give as much attention to promoting the inherent potential and productivity of the poor as is generally given to protecting the power of the privileged.

(McNamara 1981: 245)

McNamara (1981: 251) summarized the conditions of subsistence farmers in the contemporary developing world and outlined a Bank program "to increase production on small farms so that by 1985 their output will be growing at the rate of 5 per cent per year." Thus, poverty-oriented rural development projects became the hallmark of McNamara's development strategies during the first phase of his tenure as president of the World Bank (Maddux 1981).

In sum, McNamara's initiatives not only directed the Bank away from its almost exclusive concern with infrastructure (see Table 2.3), but also introduced moral perspectives that the Bank had hitherto thought best to avoid: "the whole of human history has recognized – at least in the abstract – that the rich and the powerful have a moral obligation to assist the weak and the poor" (McNamara 1981: 245). With almost missionary zeal, McNamara pressed on with his anti-poverty social lending agenda during the remainder of his presidency, foraying into sectors, such as education, employment, and nutrition, that the Bank had traditionally considered taboo. It was during these years, as part of this venture into poverty alleviation, that the Bank moved into another unfamiliar yet important terrain: funding urban development.

Chapter 3

The search for an urban agenda at the World Bank

The previous chapter identified the political and economic climate that led the Bank away from large-scale, infrastructure projects and toward social lending. Those within and outside the Bank who advocated a more socially oriented development approach generally welcomed this shift. A small minority within that camp, however, argued that the stronger concentration on rural development was misdirected, and would not succeed in adequately addressing the multifaceted socioeconomic problems that confront developing countries. This group was specifically concerned about increasing rates of urbanization in the developing world, and the resultant socioeconomic problems, which had hitherto escaped the attention of the major international development agencies. According to United Nations statistics, the aggregate increase in urban populations between 1950 and 1990 was approximately 430 million in developed countries, but 1.07 billion in the Third World – an almost threefold increase in absolute numbers, and an increase of 300 million in the 1970s alone (United Nations 1986). Owing to limitations in real income, the developing world's growing urban population created enormous problems for infrastructure and service provision, leading to the prolific growth of slums and shanty towns. By the 1970s, about 30–40 percent of urban populations in Africa, Asia, and Latin America lived in these informal settlements (Todaro 1977).

In light of this reality, Michael Cohen, who joined the World Bank in 1972 and eventually headed its urban division in the 1990s, recalled that

> A number of people took exception to McNamara's [Nairobi] speech, and said, "Have you been to Calcutta recently, Sir? Don't tell us that poverty is a problem in just rural Kenya."[1]

Cohen's statement reflects the emerging awareness among policymakers at the time that urbanization and its associated problems could no longer be ignored by international development agencies (McGee 1971; Ross 1973; Grimes and Orville 1976; Keyes and Burcoff 1976). Since cities "serve simultaneously as national and regional engines of growth, centres of technological and cultural creativity, homes of the poor and deprived, and sites and sources of environmental pollution" (Fuchs 1994: 2), understanding the dynamics of urban growth and addressing the problems associated with rapid urbanization came to be seen as crucial aspects of development policy. The United Nations and other development agencies ventured into urban poverty alleviation programs in the 1950s and 1960s. The World Bank joined them belatedly, its interest developing in the late 1960s and early 1970s as a response to the scope and intensity of urban problems in poor countries.[2]

The drastic housing shortage facing the urban poor living in informal settlements particularly caught the attention of World Bank policy-makers (Keare 1983). The Bank reasoned that, as 40–70 percent of urban dwellers were unable to afford even the lowest-cost housing provided by the public sector, they would best be assisted either by expanding the supply of low-cost housing through sites-and-services schemes or by upgrading squatter areas that contained the bulk of the urban poor's existing housing stock. Such approaches formed the core of the Bank's urban development strategy in the 1970s (World Bank 1972). However, although the Bank eventually did embrace and tackle urban poverty by addressing the housing needs of the urban poor, it was not a pioneer in this area.

The aim of this chapter is twofold: to examine the intellectual and policy trends that led the Bank to alter its once ambivalent attitude toward funding urban development programs and to overview the Bank's early urban initiatives. The chapter is divided into three sections. The first part examines theoretical and policy developments from 1940 to 1960, during the period leading up to the Bank's actual involvement in urban lending. The next section discusses the Bank's attempt to outline its own urban agenda and strategy during the 1960s and 1970s. Finally, the core features of the Bank's early urban initiatives are discussed.

Background to urban lending

During its first twenty-five years of operation, the World Bank concentrated on project-based lending for large infrastructure and industry. While a major proportion of its investments during this period were concentrated in urban areas, no actual policies existed to address the socioeconomic problems caused by rapid urbanization in the developing world. In 1960, the IDA was established to promote more flexible credits, and to make funds available for socially oriented development projects such as urban poverty reduction (discussed in Chapter 2). However, the Bank approached such lending cautiously; as late as the mid-1960s, the Bank had yet to make a loan to address the problems posed by rapid urbanization.[3]

In fact, Abrams, who studied shelter issues and served occasionally as a United Nations consultant, noted in the early 1960s that it was doubtful whether the Bank would ever do "anything significant in the field since eligible projects must have high economic priority" and the Bank's basic criterion for funding projects was productivity (Abrams 1964: 95). Housing was viewed more as a social expenditure rather than a productive investment by the Bank, at this time. According to Williams (1984: 174), the Bank's philosophy before the 1970s "polarized sectors into those which were termed 'productive' and 'consumptive.'" As a result, the Bank, "seeing housing [as] a bottomless pit" (Abrams 1964: 96), was reluctant to fund urban poverty reduction programs or housing projects in the developing world. The following comments by World Bank president Eugene Black, in an address to the United Nations Economic and Social Council, reflect this view:

> Some calculations have been made about the cost of providing houses in India during the next generation, if the population continues to grow at 2 per cent a year. If you disregard the cost of rural housing, on the somewhat optimistic assumption that it can be carried out entirely with local materials and labor, then you still have to pay for the homes of nearly 200 million extra people who, it is expected, will be living in India's cities 25 years hence . . . A sober estimate of the cost suggests that in the 30 years between 1956 and 1986 a total investment in housing of the order of 118 billion rupees, or roughly US$25 billion, will be needed. If you find a figure like that difficult to grasp, I may say that it is well over four times the total lent by the World Bank in all countries since it started business 15 years ago. Put another way, it is more than 30 times the initial resources of the International Development Association – and those resources are supposed to cover the IDA's first five years of operations.[4]

At this time, as noted earlier, the Bank's development philosophy was to finance only basic utilities and infrastructure projects, reasoning that such measures would strengthen the economies of developing societies. This, in turn, would generate the economic growth necessary for housing investment. Such a view, according to Abrams (1964: 97), meant that "little hope can be held for housing development through World Bank assistance," and that the Bank's thinking would "relegate housing problems to providence and prayer."[5]

Jacob Crane, the Assistant Director of US Public Housing, also tried unsuccessfully in 1952 to lobby the Bank to fund sites-and-services programs in Jamaica. Sir Hugh Foot, then governor of Jamaica, had invited the World Bank to send a mission to the country "to make a general economic and financial survey of the Island," and to "make an independent and objective study of the development requirements of Jamaica" (IBRD 1952: ix). Crane encouraged members of the World Bank mission also to visit Puerto Rico to examine the self-help housing schemes implemented there (R. Harris 1997a,b). In its final report, the mission argued that addressing social issues would help the economy of the country:

> In the program for developing more fully the economic potential of Jamaica, better education, better health, *and better housing* play a double role . . . .
>
> The increments which additional factories or power plants may contribute to the national product, and certainly the efficacy of comprehensive measures for soil conservation, irrigation, or the controlled use of land, will be determined by a large measure by the degree in which the people generally can understand and appreciate the objectives and can apply effective techniques.
>
> The dissemination of technical skills and general education, the improvements of nutrition, the prevention or cure of illnesses, and *the improvement of housing conditions* might, therefore, result in substantial improvements in productivity.
>
> (IBRD 1952: 115, italics added)

With specific reference to the issue of housing, the mission's report argued

> We have already suggested that 2 million pounds be spent on rural housing as an integral part of the agricultural development program. The remainder would be allocated to urban housing. Past experience has demonstrated that the construction of rental housing involves the government in substantial losses. The 2 million pounds would accomplish more if it were lent to responsible low-income families for the purpose of assisting them in building modest houses of their own.
>
> (ibid.: 127)

The mission's final report, therefore, actually advocated a more socially oriented development approach and recommended self-help housing strategies to the World Bank. The Bank, however, was not swayed; at this stage, it was skeptical of social lending in general, and regarded housing, in particular, as a non-productive investment.[6] Although it faced increasing pressure during the 1950s and 1960s to direct its lending toward social issues (as argued in Chapter 2), the Bank refused to go down this route because of its bias toward the modernization paradigm of development (see Chapter 1), which emphasized large-scale projects such as dams, railways, etc. Thus, the Bank felt that innovations to finance low-income housing ought to be left to other agencies, especially the United Nations.

During the 1950s, the United Nations began to venture into urban poverty, especially low-income housing in the developing world. In the post-World

War II period, the United Nations created a number of specialized agencies to deal with economic, social, and human rights issues. However, an agency for human settlements was not among them at this time, as international housing issues were being handled by a very small section of the Department of Social Affairs. A United Nations mission of experts that visited South Asia and South-East Asia in 1950 reported that

> with few exceptions, families in the tropics simply cannot afford to buy or rent houses built for them on a commercial basis. It is also obvious that neither governments, nor private agencies can provide housing on a subsidized basis to all in need. Practical solutions should combine the initiative and resourcefulness of the people, the rational application of local materials and skills, the mutual advantages of group work, and the best use of resources and technical knowledge available.[7]

Abrams, who was instrumental in promoting the self-help approach in the developing world, was disturbed to find that urban housing was a low priority for official aid programs as well as international aid agencies (Koenigsberger *et al.* 1980; Henderson 2000). In 1952, Ernest Weissmann, director of the United Nations' Housing, Building, and Planning Branch, asked Abrams to conduct an examination of international land problems. Abrams (1952) produced a report titled *Land Problems and Policies*, which analyzed land acquisition in general, and went on to discuss the experiences of fourteen different countries. With specific reference to shelter, Abrams argued that its lack was caused not by the shortage of land, as was conventionally assumed, but by the lack of well-conceived urban shelter policy initiatives, which exacerbated the Third World's housing crisis.

Abrams led a United Nations mission to Ghana in 1954, where he was struck by the contrast between the urban environment's modern sectors and its desperately poor periphery. He warned the Ghanaian government that their prefabricated housing projects were impractical, arguing that such projects were too costly and provided accommodation for only a tiny segment of the urban poor (Taper 1980). After reviewing the housing situation in the capital city of Accra, Abrams concluded that public housing schemes and/or mortgage loans for entire houses were not financially viable and could not adequately address the housing needs of the country. Instead, he proposed that the government need only supply or finance windows, doors, and roofs, and that the resources and skills of the population could be relied upon to complete the housing projects. Ghana's policy-makers took heed and endorsed Abrams's proposal in an early experiment in self-help housing. Abrams (1964) later adapted this proposal for Bolivia, Nigeria, and other countries.

While he believed that self-help housing could help meet low-income housing needs, Abrams (ibid.: 168) was against its promotion "as a panacea for the housing problems of the industrializing nations." He argued that, by promoting self-help housing, some public officials hoped to find an easy and cheap solution to housing needs "by shifting the onus from technology back to the individual" (ibid.: 170). Such a "bootstraps" philosophy, Abrams feared, required the "homeless to provide for themselves," often with inadequate assistance (ibid.: 174). As a remedy, Abrams suggested the concept of "core housing," which he viewed as a "major variant of the self-help technique" (ibid.: 181). "Cores" referred to the mass-produced dwelling spaces of one or two rooms that families could move into immediately and expand later. He concluded that this might be a more effective strategy for meeting low-income housing needs in the developing world.

As requests for United Nations housing assistance grew in the 1960s, a committee on housing, building, and planning was established in 1962. The United Nations Centre for Housing, Building, and Planning officially became a part of the Bureau of Social Affairs two years later. In 1972, the United Nations Conference on the Human Environment at Stockholm served as an important catalyst for subsequent developments in housing and human settlements; it recommended that a Conference on Human Settlements (Habitat I) should be held in Vancouver in 1976. The Vancouver conference resulted in the creation of the United Nations Commission for Settlements and the decision to establish Habitat, the Center for Human Settlements, in Nairobi (Weissman 1978).

In addition to the United Nations, the United States Agency for International Development (USAID) had been involved in urban activities since its creation in 1949. Between 1949 and the mid-1970s, approximately 4 percent of USAID's US$459.4 million in capital commitments had been used for urban development and related projects (Table 3.1).

The Housing Investment Guarantee Program (HIGP), which was established in 1961, became USAID's principal instrument for addressing housing related issues in the developing world. Its basic objectives were to promote local savings for long-term housing credit and the implementation of demonstration projects. Although it initially focused on creating capital-generating institutions to finance middle-income housing, by the mid-1960s, the HIGP began to assist low-income families with self-help housing.

Abrams knew that USAID's housing programs were partly driven by geo-political considerations: "After the clarification of Castro's long-term Communist aims ... housing money began to be dispensed in earnest" (Abrams 1964: 99). In fact, Abrams's own advocacy for addressing urban poverty and housing needs in the developing world contains the belief that deprivation and suffering increased the threat of communist revolution:

> There is no more fertile ground for revolutionary propaganda than the beleaguered cities of the underdeveloped nations. Misery, bitterness, and resentment in the teeming slums and squatter colonies, low wages and

long hours in the new factories, competition for jobs, and child labor, all recall the scene that made *The Communist Manifesto* an alluring document in nineteenth-century Europe.

(ibid.: 99)

The geo-political calculation that lay behind Abrams's humanitarian arguments eventually found its way into his promotion of low-income housing for the Third World. Such thinking resonated with that of Cold War decision-makers, who were more likely to support assistance for Third World housing if they believed that it would stop the spread of communism (Henderson 2000). In fact, controlling communism was the also the main reason behind Robert McNamara's effort to move the Bank toward poverty-based lending during the late 1960s, as argued in Chapter 2. Urban issues, especially housing, were beginning to get the Bank's attention as sources of political unrest.

A 1962 Senate sub-committee report on international housing programs is also transparent in this regard:

Social and political unrest and Communism are natural consequences of squalor conditions. The actions of these large masses of underprivileged and ill-housed people can wipe out all the gains from economic assistance in these countries.[8]

Thus, money began to flow from various sources into housing, a major urban sector, because it was seen as a means of "appeasing" the restless poor. USAID and other US government agencies began to get involved actively in low-income housing in the developing world. In 1961, for example, the US government and the Inter-American Development Bank together created the Social Progress Trust Fund with the aim of promoting social development projects in Latin America. As one of the main agencies approved by Congress as part of the Kennedy administration's "Alliance for Progress" (discussed in Chapter 2), this fund granted US$525 million in loans and concessional terms for projects in the "fields of land settlements, improved land uses, housing for low-income groups, community water supply, and sanitation and facilities for advanced education" (Blitzer *et al.* 1983: 110).

In addition to the United Nations and USAID, John Turner's (1963, 1965, 1972a,b) work and Janice Perlman's (1976) celebrated study of Rio de Janeiro's *favelas* had a tremendous impact on official attitudes toward the Third World's squatter settlements and non-conventional housing. Turner, who had developed his ideas while working as a professional architectural consultant in the city of Arequipa, Peru, initially assumed that the role of the professional was to organize the housing process: "Most professionals are brought up to believe that, once qualified, they know all that is necessary in order to decide what should be done for their clients" (Turner 1986: 14). However, he and his colleagues found that "anyone with a reasonably open

Table 3.1 USAID urban development assistance, 1949–70 (amounts in thousands of US dollars)

| Project type[a] | Capital grants and loans | | | | | |
| | Completed projects | | Ongoing projects | | Total | |
	No. of projects	Amount ($)	No. of projects[d]	Amount ($)	No. of projects	Amount ($)
City and regional planning	–	–	–	–	–	–
Environmental sanitation	33	54,511	15	63,233	48	117,744
Highways[b]	1	131	8	30,390	9	30,521
Housing[c]	25	71,989	10	87,558	35	159,547
Project support for housing	2	13,698	–	–	2	13,698
Potable water	2	1,832	15	66,422	17	68,254
Urban transit and traffic engineering	1	306	1	2,000	2	2,306
Totals	64	142,467	49	249,603	113	392,070

Source: Technical Assistance Completed Projects; Capital Assistance Completed Projects; Capital Assistance Projects, Office of the Controller, AID, W253, 6/30/70.

Notes
a The categories of environmental sanitation, housing, and project support for housing may include some assistance to rural areas.
b Inter-city highways.
c The housing investment guarantees are not included in this category.
d As of June 30, 1970.

mind soon learns that people, however poor and unschooled, know a great deal about their own situations and their own space, time, and energy" (ibid.: 15). Thus, Turner came to the conclusion that people knew very well not only what to build but how to build it, and revised his earlier "liberal authoritarian view that all local autonomous organizations tended to be subversive" (Turner 1972a: 138).

Turner observed that squatters managed to build their dwellings at less than half the amount charged by a contractor and, in the process, created an investment four or five times their annual incomes (Fichter et al. 1972: 241). As a result, he began to argue that housing should be viewed as a verb as well as a noun in that "housing is not just shelter, it is a process, an activity" (Turner 1972b: 122ff). As a corollary, a "house" should not be seen simply in terms of its physical characteristics (what it is); according to Turner, it should also be seen in terms of its meaning to those who occupy it (what it does). For Turner, under certain conditions, a shack may be supportive to its inhabitants while a "standard" house may be oppressive. By implication, the

| Technical assistance | | | | | | Total capital commitment | |
| Completed projects | | Ongoing projects | | Total | | | |
No. of projects	Amount ($)	No. of projects	Amount ($)	No. of Projects	Amount ($)	No. of projects	Amount ($)
2	128	1	32	3	160	3	160
70	28,661	5	1,879	75	30,540	123	148,284
–	–	–	–	–	–	9	30,521
76	11,272	21	16,159	97	27,431	132	186,978
3	74	–	–	3	74	5	13,772
1	141	9	8,719	10	8,860	27	77,114
7	315	1	402	8	717	10	3,023
159	40,591	37	27,191	196	67,782	309	459,852

material value of a house is not an adequate measure of its value to the user; human values, therefore, should be substituted for material use values.

Owing to the dynamism of housing need in relation to various factors (such as family cycle and stages of the migrant's life), large organizations (such as the state or municipality) always had to standardize procedures and products. As a result, housing policies formulated by these organizations mostly missed the changing needs and priorities of individuals. Therefore, Turner believed that the main components of the housing process ought to be left to individual users. While he did not advocate that dwellers should build their own housing without state help, he did argue that they should determine their own needs individually through decentralized, local institutions. In sum, Turner's main contention was that neither the commercially motivated private sector nor the politically (and sometimes commercially) motivated public sector ought to govern the building of houses. Housing, in Turner's view, ought to be left to the "popular sector," which could make ample use of the plentiful, renewable, and partly non-monetary resources available to them. This would be the main advantage of building through the popular sector. Turner cited the economic advantages of popular sector housing as follows:

> The bureaucratic, heteronomous system produces things of a high standard, at great cost, and of dubious value, while the autonomous system produces things of extremely varied standard, but at low cost and high use value. In the longer run, the productivity of centrally administered systems diminishes as it consumes capital resources, while the productivity of locally self-governing systems increases as it generates capital through the investment of income.
>
> (Turner 1976: 82)

One of the key policies resulting from Turner's work was the recommendation that governments should stop trying to provide standard housing for the poor, and instead use the human potential of the low-income population by permitting and enabling them to house themselves.

Janice Perlman's (1971, 1976) seminal work has seriously challenged the "myth of marginality" of the urban poor. Her ideas were also critical in shaping the debate among development institutions on the need for urban lending. Perlman divided her analysis into four component parts – social, cultural, economic, and political – and, in each instance, successfully illustrated how squatters were integrated into mainstream urban life. According to Perlman (1976: 2), the apparent marginality of the squatter's way of life is a function not so much of its segregation as of its exploited integration. Squatters, she wrote,

> are not economically marginal but exploited, not socially marginal but rejected, not culturally marginal but stigmatized, not politically marginal but manipulated and repressed.
>
> (ibid.)

Perlman (1980: 251) cogently argued that the perception of squatters as a "parasitic drain on the urban economy . . . [who] suffer from all forms of social disorganization" is false and that squatters are actually integrated into "mainstream" society, but in a manner detrimental to their own interests.

The studies of Perlman and Turner were decisive in easing some official antipathy toward squatter settlements in the Third World city. They resulted in the formulation of some new policies that sought to incorporate both the capabilities and the needs of the poor. This shift in thinking later prompted international development agencies, and eventually the World Bank, to lend toward low-income housing provision as a way of getting involved in the strategically vital urban sphere. Table 3.2 highlights some important developments that impacted housing policy toward the urban poor.

By the 1970s, governments and international development agencies

Table 3.2 Influential international events in low-income housing, 1930s to 1972

1930s	1950s	1960s	1972
Jacob L. Crane, assistant director of the United States Public Housing Department, promotes the term "aided self-help housing" among planning officials in the US and in developing countries	Charles Abrams promotes "core housing" through the United Nations and argues that the lack of housing in developing countries is caused not by a shortage of land but by ill-conceived policies	John Turner argues that housing should be viewed as a "verb" and that neither the commercially motivated private sector nor the politically motivated public sector ought to govern house-building, which ought to be left to the "popular sector" USAID and other US government agencies get involved in low-income housing in order to reduce the appeal of communism among the urban poor of developing countries	First World Bank sites-and-services program in Senegal

increasingly came to accept squatter settlements as an inevitable consequence of Third World urbanization.

For reasons discussed in Chapter 2, and the housing debate discussed above, the Bank was forced to re-examine its previously reluctant stance on urban poverty reduction and began to invest substantial resources in sites-and-services and squatter upgrading schemes. As I observed in the previous two chapters, the Bank is not always a pioneer of major shifts in thinking (indeed, the Bank is often late to join ongoing debates), but it has the power to appropriate key aspects of the debates, inflect them to suit its own agendas, and endorse its positions such that they become the new orthodoxy. The Bank's conclusions eventually become the official, conventional, or commonsense views such that everyone else follows suit until a new debate arises. For example, self-help housing became the new orthodoxy as the Bank adopted and promoted these schemes as affordable, feasible solutions to the developing world's housing crisis. The circumstances leading to the Bank's embrace of self-help strategies and the core features of its urban program are discussed below.

In search of an urban agenda

The 1960s were a period of critical examination in which the dominant models of development were questioned by the international aid community. Although some Third World countries recorded impressive rates of economic growth, the increasing poverty and inequality that characterized much of the developing world caused many development practitioners and theorists to argue that the goals of fulfilling basic humans needs and alleviating poverty should share the same high priority given to economic growth (ILO 1976; Srinivasan 1977; Streeten 1977; Rimmer 1981; Streeten et al. 1981). By the early 1960s, the World Bank had begun to extend the scope of its lending program into the social field. McNamara accelerated this trend, greatly expanding lending into the areas of agriculture and education during the late 1960s and early 1970s (for reasons discussed in Chapter 2; see also Table 2.3). However, although the Bank had committed itself to addressing rural poverty, it still lacked, at this stage, a coherent strategy to deal with the problems posed by rapid Third World urbanization.

M.A. Cohen (1983: 3) points out that although there was a reluctance on the part of the international community to address urban issues, owing to the belief that "urban investment . . . would direct much needed resources away from the rural sector," individuals within and outside the Bank gradually began to realize and argue that a focus on rural development alone would not address the multiple manifestations of Third World poverty, much of which is urban in nature. For instance, Edward Jaycox, first Director of the World Bank's Urban Division,[9] noted that:

Even with much greater efforts in rural development, however, there are limits to what can be achieved. The amount of land that realistically can be brought under cultivation in most countries is either quite limited or can only become productive at high and increasing costs – for clearance, infrastructure, irrigation, and settlement. Furthermore, even with greatly increased output per hectare, which assumes the continuing spread of technology, fundamental questions remain. How much labor can be productively absorbed in the development process? Can *per capita* rural incomes be raised to reasonable levels without the substantial exodus of people from the land?

(Jaycox 1978: 11)

As the Bank realized that poverty alleviation could not be limited to the rural sphere, it sought an effective national strategy to address poverty and socioeconomic inequality that would not only continue and increase the gains of rural productivity, but would also address the challenges posed by rapid urbanization. Urban poverty, together with the gross inefficiencies and inequities that characterize urbanization in the developing world, would have to be a part of any agenda that addressed Third World poverty (World Bank 1975a). While some Bank policy-makers realized that the institution would have to venture into the field of urban development for strategic political and humanitarian reasons, they were still uncertain about what the appropriate policy response and intervention might be.

The formulation of this response began in two separate bureaucratic locations at the World Bank: in the Development Economics Department, headed by David Henderson,[10] and in the Special Projects Division, headed by Robert Sadove[11] (Figure 3.1).

Kenneth Bohr, a civil engineer based in the Special Projects Division in 1971, recalled that "for many who were concerned with poverty alleviation, it was obvious that the urban problems could not be ignored."[12] The difficulty, according to Bohr, lay in formulating a strategy that was consistent with the Bank's conservative fiscal policy. In a background briefing paper, Bohr wrote:

We are aware of the deficiencies in service and facilities in the large cities of developing countries ... What we don't have is some quantitative estimate of these various situations that might provide a basis for a comparison [of] problems and cities and a realistic evaluation of possibilities. The ability to present urban problems in some concreteness is the only way to get across to the management and the Board that we are making some headway and are at least beginning to size up our problems in operational terms.[13]

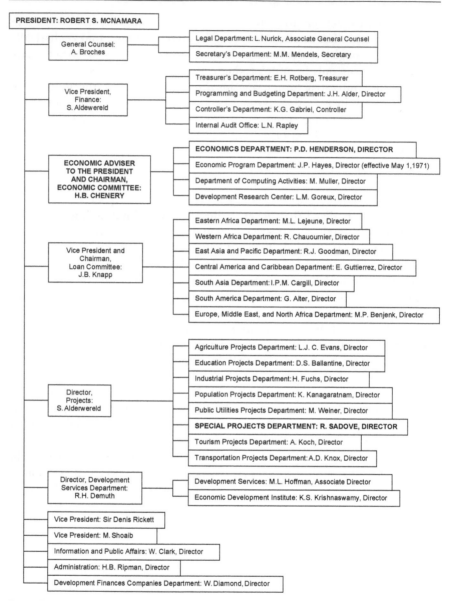

Figure 3.1 World Bank organizational chart, 1971.

Thus, although certain sections of the Bank recognized that an urban strategy was necessary, some confusion remained regarding an appropriate direction. In a Bank memo schematically outlining a possible approach to urban lending, Bohr acknowledged that the Bank would be "entering fields where our own experience is limited and a growth of competence will be required on the lender's as well as on the borrower's part."[14] While the memo argued that "the effective operation on urban problems should start from a consideration of an urban area in a similar manner to the Bank's usual consideration of a country – the economic prospect of the city, policies for its growth and administration, its investment program, financial viability, etc.," Bohr did acknowledge that "the analogy should not be carried too far."[15] The new focus on rapid urbanization in developing countries would "amount to a considerable change – new problems, new ways of looking at old problems, and just plain old problems, long recognized but no longer possible to avoid."[16]

In 1972, the Special Projects Division produced a sector paper that identified the scale and nature of the problems posed by rapid urbanization in developing countries (World Bank 1972: 53–5). The paper mentioned that "the proliferation of squatter settlements and slums and the rising backlog in urban services" led to the realization that "development implies much more than an expansion of output" (ibid.). Also, in spite of the fact that a large part of the Bank's previous lending had been urban-oriented, the paper pointed out that emphasis on large scale infrastructure projects "restricted the types of the Bank's urban projects" (ibid.); these, according to the paper, would now have to be re-evaluated.

> With problems of urbanization becoming increasingly severe, attention is now being devoted to how the Bank's operations can be more consciously and effectively related to improving the efficiency of the urban centers, both for production and for living.
>
> (ibid.)

The paper also acknowledged that the Bank was confronted by a "shortage of experience and expertise on urbanization problems" (ibid.). According to Bohr, one of the main purposes of this paper was "to get Bank operations to recognize the urban economy as a unit." Bohr felt this would be a worthwhile endeavor because it "would be helpful for discussing investment choices that bore on poverty and employment and for a discussion on the relationship between urban and rural economies."[17] Additionally, the 1972 sector paper contributed to the Bank's emerging strategy for dealing with the urban problems by advocating sites-and-services and mass transit, as well as simpler and cheaper provision of power, water, and sewerage.

This urbanization sector paper was released to coincide with the presentation of the Bank's first urban sites-and-services project (to be implemented in Senegal) to the board of governors. However, at this time, the Bank had not yet articulated a clear urban agenda for itself. Many board members were skeptical of the Bank's social involvement in general, and of the sites-and-services approach in particular. Some members of the board argued that the housing problem was so large that "the Bank could not hope to make a dent in it."[18]

A parallel program on urban research was launched within the Development Economics Department (DED) of the Bank, headed by Douglas Keare.[19] Keare began working at the Bank in 1967 as its representative in East Pakistan. In 1969, he was invited to work in the DED on urban issues.

> When I was in Dhaka, David Henderson[20] asked me if I had any interest in taking over a division which was not getting off to a good start. I said no initially because I had just relocated to Dhaka. When the Civil War came along and relocated me, I said yes. The job was still open. We renamed it the Urban and Regional Economics Division. I got started setting up a new urban division in September of 1971. At that time, there was another division within the Special Projects Department. [Author's note: This is the Special Projects Department headed by Robert Sadove, mentioned above.][21]

While the Special Projects Division concentrated on the new sites-and-services loan to Senegal and getting new proposals ready for board approval, the urban research section of the DED devoted its resources and energy to studying the process of urbanization itself. The Department's top priorities included housing needs, land markets and policies, urban public finance and administration, rural–urban migration, and growing urban poverty. In 1975, under Keare's direction, the DED produced an important document, *Housing: Sector Policy Paper* (World Bank 1975b).[22] One of the main purposes of the paper, according to Keare, was to address the fear that the Bank was subsidizing consumption by lending toward sites-and-services schemes and squatter upgrading.[23]

By advocating for sites-and-services and squatter upgrading, the Bank aimed to reconcile two seemingly contradictory goals: poverty alleviation and economic productivity. On one hand, Keare, Jaycox, Cohen, and other policy-makers at the Bank tried to convince Third World governments and policy-makers of the need to adopt sites-and-services and squatter upgrading as a cost-effective, graduated approach to housing the urban poor. On the other hand, they had to reassure mainstream economists within the Bank that funding housing development was more than mere "social work," and that such a venture would indeed contribute to economic growth. To this end, Keare recalled, "We were at great pains to show that housing was a productive investment, that it was a tool for macro-economic development."[24] The DED housing sector paper was written specifically to address this issue. The multiplier linkages of housing in the urban economy could be substantial, Keare and others argued, and could contribute to higher national productivity by making the underutilized labor, material, and financial resources productive. However, although urban work had begun at the Bank around the late 1960s, "the urban troops were not fully recruited or very well organized by the time of McNamara's Nairobi address."[25] While a small group of urbanists within the Bank were able to gain approval for and implement projects such as the sites-and-services scheme in Senegal, they were unable at this stage to convince the Bank's senior management to address urban problems as part of an overall poverty alleviation agenda.

In the 1970s, the annual addresses of the World Bank's presidents set the tone for new agendas, as the World Bank did not publish world development reports then. In fact, these addresses were the principal vehicles for delivering the Bank's statement of policy to the development community. In the contemporary period, the world development reports have replaced presidential addresses as indicators of the Bank's thinking on various issues. Therefore, after McNamara's Nairobi address, the urbanists at the Bank not only sought to formulate a coherent urban poverty alleviation strategy for the Bank, but also lobbied at the presidential level to make such a strategy central to the Bank's poverty agenda. This idea was received with some skepticism in certain quarters. According to Jaycox, the first director of the Bank's Urban Department, "Some people in the Bank were making jokes that next we are going to have suburban development, or an outer space development program."[26] Peter Cargill, senior vice-president for finance, wrote in a memo,

> I am surprised that urban poverty should be regarded as an important topic for McNamara's speech. None of these social problems, including this one, can really be resolved except in the context of economies which have a reasonable rate of growth.[27]

This kind of skepticism did not deter the urban advocates, according to Jaycox, who labored on to articulate an appropriate urban policy response for the Bank.[28] In 1973, his Urban Department merged with the Department of Transportation and became known as the "Department of Transportation and Urban Development." Jaycox, who became the head of the new department, recalled that it lacked a coherent agenda when it was formed: "They had no idea what they were supposed to do. They didn't have any focus at all."[29] Behind that criticism was the fact that the newly created department faced numerous bureaucratic challenges in designing appropriate programs during a period of great organizational change within the Bank and political change outside of it. Jaycox observed that those who were formulating an urban policy for the Bank searched for "the urban equivalent of the small farmer – that is, a targetable population that could be the recipient or direct beneficiaries of productive investments, not simply welfare transfers."[30] According to Douglas Keare, Jaycox's appointment energized the new department because he was one of "the main movers in shaping the Bank's early urban agenda."[31]

In 1974, the two departments dealing with urban issues, the Urban and Regional Economics Department under Keare and the Department of Transportation and Urban Development under Jaycox, coordinated the

Urban Poverty Task Force to investigate the nature and extent of urban poverty, as well as how the Bank might respond to it. Under the direction of Michael Cohen, the Task Force produced two major reports and played an important role in formulating the Bank's urban policy (World Bank 1975a, 1976; Beier *et al.* 1976).

The Task Force's first report, titled *The Task Ahead for the Cities of the Developing Countries*, set out to document the past and projected patterns of future urban development in Third World countries (World Bank 1975a). Examining such indicators as the growth, size, and distribution of urban areas, inequity and poverty in the city, and the absorptive capacity of cities, the report proposed policies to address these challenges. Urbanization could be dealt with successfully, the Task Force stated, if national and local governments recognized and addressed the challenges it posed. This report identified the "low productivity" of the poor as the major hurdle facing the urban economies of developing nations, and recommended two main points of intervention: first, policies should be aimed at increasing the demand for labor and/or upgrading its quality; and, second, the quantity and quality of services, particularly public services for the urban poor, should be improved within existing resource constraints (Beier *et al.* 1976). In addition to the

Table 3.3 Urban Poverty Task Force policy recommendations

Country type	Policy recommendations
Type I	
Highly urbanized and incomes in urban areas relatively high, with only a small percentage of poor urban dwellers	Improve distortions in the labor market caused by various government regulations Promote the informal sector Security and legality of tenure to replace squatting Upgrade family planning efforts
Type II	
Over half of the population live in rural areas	Improve physical infrastructure Promote export-oriented industrial development Squatter and slum improvements
Type III	
Predominantly rural but urbanizing rapidly	Similar proposals to type I/II Reduce the public sector
Type IV	
Dominated by severe pressures on the land in largely rural societies with a subsistence level of income. Only about 20 percent of the population live in urban areas but rates of urban growth are high	Promote small labor-intensive enterprises Address basic needs to squatters, such as drainage, supply of water, etc. Family planning and education to enable younger generation to participate in the urban economy

Source: Beier *et al.* (1976).

general policy recommendations, the report proposed recommendations for a typology of four different patterns of urbanization in the developing world (Table 3.3).

The second publication of the Task Force was the *Urban Poverty Action Program* (World Bank 1976). Critical of the Bank's previous urban work, this report concluded that none of the Bank's previous urban lending focused on socioeconomic inequality and poverty:

> The review of past lending has highlighted the lack of any systematic attention in project appraisal or program development to urban absorption [of the labor force], income distribution, or employment characteristics.
>
> (ibid.: 12)

The report found that only one-third of the urban projects for the fiscal years 1973–5 "provided clear evidence of [generating] substantial unskilled employment" (ibid.: 3). Less than a quarter of the projects could be said to have favorable impacts for the urban poor in improving their relative access to urban services; some positive impact on urban institutions and/or policies could be deduced in only a third of the projects (ibid.: 5). The report also found a "concentration on large infrastructure and industrial projects" that provided "little evidence of direct benefits to the poor or of direct increases in the capacity of cities to absorb the population growth" (ibid.: 7). Reflecting on the Bank's previous concentration on large-scale infrastructure project lending, Jaycox also noted that

> scarce capital is being concentrated on relatively few workers, increasing their productivity greatly, but leaving most of the workforce without access to capital and with very low productivity. The Bank's traditional operations in manufacturing and mining adhere to this general pattern ... [The Bank] must find ways of reaching the poor directly in areas that it has hitherto not touched – such as service enterprises, small-scale operations, the self-employed artisans, and cottage industries.
>
> (Jaycox 1978: 12)

The Task Force's analysis concluded that the Bank's urban programs should have a more direct impact on the approximately 190 million people living in absolute poverty in urban areas of the developing world. The Task Force recognized that the Bank's urban programs would have to be multifaceted, unlike its rural programs, which targeted a "fairly homogeneous group of producers with access to the basic factors of their production process" (World Bank 1975a). Furthermore, it noted that "the multidimensional characteristics of urban poverty did not lend itself to a single urban strategy for the Bank," and that "the Bank's programs ought to have a positive impact on the "absolute poor of the developing world's urban areas" (ibid.).

The Task Force recommended a number of goals for the Bank's urban lending program. With respect to shelter, the Task Force sought to address, by 1980, one-third of the annual increase in the unserviced, informally settled populations. Regarding the water supply, it aimed to ensure that all people within a given project area should have access to safe water. Calling for a "sectoral shift" in the Bank's lending, the report stressed that the Bank ought to move away from its traditional large-scale infrastructure projects in order to focus on low-income housing, sanitation, and water provision as sectoral programs.

In 1973, in addition to the work of the Urban Poverty Task Force, Jaycox sponsored a major international conference at the World Bank to deliberate upon appropriate urban policy. Bringing together some of the major thinkers from around the world, the three-day meeting generated much interesting discussion on the nature and challenges of Third World urbanization. It did not, however, result in concrete policy responses. Jaycox wryly recalls that:

> They [conference participants] were not a hell of a lot of help, we were all over the place. People wanted to worry about land, about taxation and management, and resource mobilization, master planning, and all that. This was expected, given the fact that they were all gurus in various parts of the picture. We were still left with the challenge of articulating appropriate policies to address the problems posed by urbanization.[32]

The participants concurred that the pathologies of Third World urbanization would have to be addressed, but little else. On the last day of the gathering, Jaycox expressed his disappointment to Otto Koenigsberger over lunch: "What the hell am I going to do? I have all these people and we have to finish after lunch!"[33] Koenigsberger, according to Jaycox, agreed that

> what we call the urban problem is the lack of infrastructure that is affordable and can service the poor. They [the poor] are being impoverished by the lack of services, infrastructure and housing, and secure places to settle. Why don't we tackle that problem directly because it is a symptom and also a cause of poverty in Third World cities?[34]

Together they presented this idea at the final session of the conference and it was unanimously accepted as an appropriate policy direction for the Bank. Thus, the deliberations of the 1973 conference, the discussions of the Urban Poverty Task Force, and a series of policy papers (World Bank 1972, 1974, 1975a,b) formed the basis of McNamara's major policy speech on urbanization in 1975.

Whereas McNamara's 1973 "Nairobi address" was decisive for rural development and poverty alleviation, his 1975 speech at the Board of Governors meeting in Washington, DC, was the turning point for urban development, according to Cohen,[35] who was Urban Task Force Coordinator at the time. In this address, McNamara (1981: 295–334) committed the Bank to a major undertaking to help national governments alleviate poverty in their cities. Outlining the reasons for the Bank's offensive against urban poverty, McNamara referred to two basic facts: that the proportion of the developing countries' population living in cities would increase greatly as the century came to a close and that over 1.1 billion people, most of them poor, would be absorbed by cities of the developing world. McNamara acknowledged that the challenges of urbanization in the developing world are "more complex than the problem of poverty in the countryside" (ibid.). Nevertheless, he felt that the Bank was ready to undertake "comprehensive efforts" to help governments deal with rapid urbanization (ibid.). Along with humanitarian reasons for addressing the "unspeakable grim life" of the urban poor, McNamara also identified urban poverty as a problem that "did not favor political delay" (ibid.). Emphasizing the links between political instability and national security (discussed in Chapter 2), McNamara worried that:

> An even more ominous implication is what the penalties of failure may be. Historically, violence and upheaval are more common in cities than in the countryside. Frustrations that fester among the urban poor are readily exploited by political extremists. If cities do not begin to deal more constructively with poverty, poverty may begin to deal more destructively with cities.
>
> (ibid.: 316)

He then proceeded to outline four broad steps for addressing urban poverty:

1   increasing earning opportunities in the informal sector;
2   creating more jobs in the modern sector;
3   providing equitable access to public utilities, transport, education, and health services;
4   establishing realistic housing policies.

McNamara argued that, as existing public housing schemes in the developing world did not reach the urban poor, the proliferation of slums and squatter settlements was the inevitable result. Authorities had generally disapproved of these unsightly and unsanitary structures, frequently deploying demolition as official urban policy. Such measures, however, did not deal with the realities confronting the cities of the developing world, for McNamara:

too often cities have failed to find any solution short of demolition to deal with them. The fact is that upgrading squatter settlements can be a low-cost and practical approach to low-income shelter. Upgrading legalizes the settlement, and provides security of tenure.

(ibid.: 327)

McNamara passionately implored planners and municipal bureaucrats not to regard the urban poor as a "statistical inconvenience," and urged development practitioners to confront urban poverty seriously. He argued,

Cities exist as an expression of man's attempts to achieve his potential. It is poverty that pollutes that promise. It is the task of development to restore it.

(ibid.: 321)

With that speech, McNamara committed the Bank to a course of urban poverty alleviation. His urban program had two basic goals: to create productive non-farm employment opportunities at much lower capital costs per job and in much greater numbers for the urban poor; and to develop urban programs to deliver basic services at standards that were affordable to poor urban residents. Housing received special emphasis in McNamara's speech, especially sites-and-services schemes and the upgrading of existing squatter settlements. In 1976, in order to effectively coordinate these new urban initiatives, the Urban Department was uncoupled from the Transportation Department and headed by Edward Jaycox.

Early urban initiatives

McNamara's policy speech and the World Bank's urban initiatives during the early 1970s were influenced by six general factors. First was the realization that policy-makers could no longer neglect the increasing rate of urban growth, and the expansion of slum and squatter settlements in particular. Between the late 1960s and early 1970s, World Bank studies of sixty-six major cities in forty-three developing countries found that in 58 percent of the cities more than a third of the inhabitants were living in squatter settlements. In 30 percent of the cities, more than half of the population were living in informal settlements (Grimes and Orville 1976: 116–17). Second, pioneering work by the United Nations, Crane, Abrams, Turner, Mangin, Perlman, and others extolled the capabilities of low-income people to build their own housing and procure basic services despite neglect and opposition from national and local governments. Their studies also demonstrated that the urban poor were central, rather than marginal, to the urban political economy. Third, it became apparent that the income gap between rich and poor in the developing world did not simply coincide with the divide between rich urban centers and poor rural areas; there were massive disparities within urban areas themselves. In light of this reality, the fourth reason for the Bank's engagement in the urban sphere was the realization that conventional economic policies were inadequate for dealing with the increasing number of poor, for whom the benefits of economic growth did not "trickle down."

Fifth, growing concern with the productivity and welfare of the poor led to a new strategy for economic development; termed "redistribution with growth," this strategy attempted to alleviate absolute poverty by channeling some of the benefits of economic growth toward the goal of enhancing the urban poor's productivity (Chenery *et al.* 1974). For example, Mahbub ul Haq (introduced in the previous chapter), senior economic policy advisor at the Bank under McNamara, stressed that

> developing countries should define a minimum bundle of goods and services that must be provided to the common man to eliminate the worst manifestations of poverty: minimum nutrition, educational health, and housing standards.
>
> (ul Haq 1976: 34)

At the time, such thinking paralleled the basic needs approach being promoted by agencies of the United Nations and individuals like Seers (discussed in Chapter 2); during the 1970s, as the Bank tried to demonstrate that meeting the poor's needs did not involve any long-term trade-off with economic growth (Streeten *et al.* 1981), the "redistribution with growth" idea was being incorporated into the Bank's overall strategy, as discussed in Chapter 2. Thus, the Bank's urban lending policies attempted to meet basic needs within a framework of continued economic growth, while increasing the earning capacity of the poor. Finally, geo-political considerations pushed the Bank into funding urban projects in the 1970s; this was reflected in McNamara's 1975 speech on urban development to the World Bank board of governors. Behind his words of caution that failure to intervene could foster civil upheaval and violence was the unstated premise that the urban poor could undermine economic and social stability if they were denied access to better housing and urban facilities (Stren 1978: 3).

The stated objective of the Bank's urban intervention was "to assist member governments to develop approaches for the efficient and equitable provision of urban services and employment," and to ensure that these investments address the needs of the poor, who constitute the majority of the urban population in most developing countries (M.A. Cohen 1983: 3). The key aims of the Bank's urban programs in the 1970s, according to M.A. Cohen (ibid.), were to:

• provide low-cost technical solutions for shelter, infrastructure, and transport, which the urban population could afford and which could be improved over time;

- demonstrate that it was possible to provide services for most of the urban poor on a non-subsidized basis;
- illustrate the feasibility of comprehensive urban planning and investment procedures suitable to rapidly changing urban conditions;
- demonstrate the replicability of projects incorporating these objectives, i.e. the ability of such projects to be self-financing and self-sustaining, and able to be extended or reproduced elsewhere.

In response to the variety of concerns facing the urban sector, the Bank engaged in four principal types of project lending for urban development in the 1970s: projects for shelter, projects for transportation, integrated urban projects, and regional development projects.[36] The preliminary framework for the Bank's urban policies was advanced in three key publications: *Urbanization (Sector Paper)* (World Bank 1972), *Sites and Services Projects* (World Bank 1974), and *Housing: Sector Policy Paper* (World Bank 1975b). Although these publications addressed a host of issues pertaining to housing and urban economics, they developed a pragmatic trinity of criteria that would guide future World Bank urban policy in general: affordability, cost recovery, and replicability.

"Affordability" became one of the Bank's catch-phrases in its low-income housing proposals for the Third World. In contrast to projects conducted by the heavily subsidized public sector, one of the primary aims of the Bank's projects was to make housing affordable to low-income households, but without the payment of subsidies. The Bank felt that budgetary limits, rather than professionally designed housing standards, ought to determine methods of construction and housing standards. "Cost recovery," another favorite term, reinforced affordability as a means of avoiding the self-perpetuating expansion of unaffordable government subsidies in the budgets of developing countries. By holding finance capital intact, cost recovery was said to promote the "replicability" of projects, and eventually eliminate squatter settlements altogether.

Echoing the views of John Turner and others, the Bank argued that conventional permanent housing was not possible in developing societies, given the limited resources available to the public sector.

> At present income levels, it is impossible for most urban inhabitants to afford even minimum standards of conventional permanent housing. Given the limited resources available, there is no prospect of adequate provision of subsidized housing for new additions to the population.
>
> (World Bank 1972: 5)

Explaining the housing deficit in market terms, the Bank argued that there were "substantial gaps between housing supply and demand in most cities of the developing world" (World Bank 1974: 4). As long as Third

World national policy-makers remained committed to the idea of providing conventional housing for the urban poor, they would not, according to the Bank, be able to meet the enormous existing demand of their cities. To deal with this situation, the only viable method the Bank could identify was to lower the cost of housing to make it accessible to the urban poor, who were excluded from the "official" housing market. The only way to make housing accessible, in the Bank's view, was to lower standards. (See Chapter 5 for a discussion of this issue in Zimbabwe.)

The Bank began to argue that housing costs could be lowered substantially if standards were reduced, claiming that existing housing standards in the developing world were unrealistically high. By reducing costs, the Bank aimed to remove barriers preventing the poor from building legal permanent houses, as argued by Cohen:

> It was realized that urban shelter and infrastructure programs on a scale required for a country to meet its basic needs would continue to exceed the resources available in that country – unless the shelter programs could be undertaken at standards low enough to be affordable by the beneficiary population. Only in this way could self-sustaining, large scale programs be launched.
>
> (M.A. Cohen 1983: 9)

The Bank expected this policy reorientation to benefit the majority of the urban population living in unserviced slums and squatter settlements, rather than the urban middle-classes, who typically benefited from conventional housing programs.

The sites-and-services approach challenged the conventional wisdom of the time by arguing that households ought to be able to build houses according to their own preferences in design, materials, and schedule. Senior World Bank urban economists such as Mayo argued that Third World policy-makers refused to recognize that "slum housing represents a large part of the poor's capital stock; destroying capital is not a good prescription for development" (Mayo *et al.* 1986: 184). Policies that improve the conditions of squatter settlements were thought to be cheaper and more effective in the long run.

A major problem that faced sites-and-services and squatter upgrading schemes was that the governments of recipient countries had great difficulty in accepting them as solutions to their housing crises. Third World governments wanted to modernize urban areas and saw informal settlements as an eyesore. In this regard, Bank officials encountered substantial resistance from policy-makers in the developing world as they tried to promote this idea. Kenneth Bohr, who was involved in the Bank's early urban initiatives, notes that "most governments argued, on political and technical grounds, that reduced standards were unacceptable because they were not good enough for our people."[37]

However, this ambivalence was overcome, according to Bank officials, when the initial projects in both sites-and-services and slum upgrading proved to be feasible alternatives to the continued growth of uncontrolled, unserviced, settlements. Officials point to a number of examples to illustrate this point. In Nigeria, for example, the cheapest house financed by the public sector and built to conventional standards cost US$40,000 in 1978. By contrast, in the same year, a shelter unit financed by the Bank's first project in Nigeria cost less than 10 per cent of the conventional housing project (Onibokun *et al.* 1989). Similar examples are cited from India and elsewhere (World Bank 1994a). Thus, from the Bank's perspective, policies on design standards changed in many countries when governments realized that it was possible to provide acceptable shelter and infrastructure at affordable costs per unit without public subsidies. Douglas Keare summed up that "resistance from policy makers to lowering building and housing standards was countered by reality."[38] Thus, Bank officials and reports took credit for Bank-introduced design standards that lowered shelter costs, dramatically in some cases. The case of Zambia is often cited, where complete houses in sites-and-services projects cost less than one-fifth the price of the least expensive government-subsidized housing. In El Salvador also, the better-quality sites-and-services project houses were less than half the cost of the cheapest conventional house (Keare and Parris 1982).

Since governments in the developing world had no effective methods of their own to address the problems of urbanization, they increasingly embraced the ideas promoted by the World Bank. Meeting the shelter needs of the urban poor offered the Bank a "strategic point of entry into the urban scene," according to Cohen.[39] While they hesitated to accept the Bank's housing prescriptions, Cohen perceived that "most governments could not afford the financial costs of conventional housing solutions nor the political costs of bulldozing existing squatter settlements."[40] Pugh (1994: 177) further observed that in addition to the power of persuasion, the Bank was able to enforce its policies though financial power, which allowed it to bridge gaps between theory and practice in whichever manner it saw fit.

Sites-and-services schemes

In June 1972, the executive directors of the World Bank finally approved the first "urban development" loans for shelter programs in Senegal, providing US$8 million in IDA credit for sites-and-services projects. These original projects were followed in 1973 by loans for sites-and-services projects in Calcutta, Managua, and Francistown (Botswana), among other cities. In successive years, new loans were approved for sites-and-services projects in San Salvador, Jakarta, Nairobi, Lusaka, Dar-es-Salaam, Manila, Kuala Lumpur, Madras, Abidjan, La Paz, Rabat, Cairo, and Alexandria, among other major metropolises, as well as smaller cities in various countries (Blitzer *et al.* 1983).

Between 1972 and June 1981, sixty-two projects amounting to US$2 billion were approved, including projects for shelter, urban transportation, integrated urban development, and regional development. The sectoral distribution of these programs is depicted in Figure 3.2.

During this period, about 60 percent of the Bank's urban interventions took the form of shelter programs and were heavily concentrated in sub-Saharan Africa. The Bank estimates that it was able to help twenty-nine developing countries provide approximately 310,000 lots through the sites-and-services approach, and improve some 780,000 lots through upgrading efforts. Assuming that there were ten people per lot, the Bank's shelter projects claimed to benefit some 10 million people by the early 1980s (Baum and Tolbert 1985: 296).

A project in Zambia, for example, enabled the preparation and servicing of 4,400 residential plots in six sites. In Egypt, about 4,600 service plots were set up for a population of approximately 23,000 people (Adegunleye 1987). A project in Tanzania prepared about 19,000 surveyed plots to be allocated to low-income applicants in five towns. In Thailand, a project offered sites-and-services in Bangkok for about 3,000 housing units (Laquian 1983a), while a project in Calcutta provided residential lots and rental units for a target population of about 45,400 people (Pugh 1988, 1989a,b). About 13,470 plots were serviced and 9,760 core units were constructed in Madras, India, through a Bank project (Pugh 1988, 1989a,b); in Latin America, a project in Colombia provided 7,300 sites in secondary cities. A project in La Paz, Bolivia, developed an estimated 5,525 serviced sites with core dwelling units,

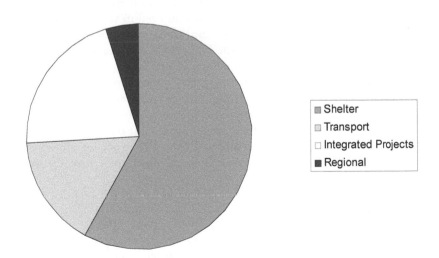

Figure 3.2 World Bank urban projects by type, 1972–81. Source: Mason and Asher (1973: 876).

while Bank funding in El Salvador allowed for about 7,000 serviced lots and 3,500 basic dwellings (Ayres 1983; World Bank 1985a).

The types of shelter within sites-and-services schemes varied. Projects ranged from surveyed plots only (the cheapest option, as in the first Tanzanian project), through plots with sanitary cores, to plots containing various housing options. Options differed according to the size of plot, size of basic dwelling, and the balance of community versus individual services. Depending on the array of options available in a given project, the social composition of the sites also varied. Where only a surveyed plot or a plot with a sanitary core was provided, there was little heterogeneity among site occupants (as in Tanzania); where there were diverse offerings, there was more heterogeneity (as in a project in Côte d'Ivoire). The projects were also built through different means. Some relied on self-help or mutual help, while others hired contract labor for site preparation and house construction (M.A. Cohen 1983; Laquian 1983a; World Bank 1985a).

The sites-and-services schemes posed a series of challenges for the Bank. The government had to acquire land for projects from private holders, which frequently led to delays and difficulties. Incompatibility between projects, existing zoning regulations, and national building codes often contributed to tensions between Bank officials and national policy-makers. Furthermore, sites-and-services schemes did not fully appreciate locational factors. Most of the projects were located in the urban periphery, requiring a relocation of inner-city squatters that often resulted in political, economic, and social dislocation. For example, the project in Senegal was located in the periphery of the city where land was cheaper; planners had hoped that introducing services into these areas would influence the direction and pace of the city's growth. However, one of the difficulties encountered in this project was the fact that it disrupted people's lives by forcing them to move far away from their places of work (Laquian 1983a). In the Philippines, for example, 80 percent of the relocated people left the project areas within two years to return to metropolitan Manila. Overloaded transportation systems were not able to serve peripheral urban areas. Sites-and-services schemes often overlooked the fact that, for squatters, locational choice was often governed by such factors as proximity to jobs, entertainment, and educational opportunities for children. Uprooting squatters and slum dwellers from locations where these resources were available wreaked havoc upon their already fragile existence (Laquian 1977, 1983b).

Squatter upgrading

In response to criticism that sites-and-services schemes were not reaching the poorest urban residents, subsequent urban projects of the Bank began to incorporate more slum renovation strategies. In Tanzania, slum upgrading accounted for only 16 percent of total project costs for the first group of proj-ects. However, in the second group of projects, approved about three years later (1974–8), the percentage of the total costs accounted for by slum up-grading doubled, or rose to 32 percent. In Calcutta's first project, approved in August 1973, the slum upgrading component accounted for only 3 percent of project costs; by the second project, approved in December 1977, this compo-nent had risen to 18 percent of project costs. Similar trends were evident in projects in Kenya as well (Keare and Parris 1982; Laquian 1983b; Mghweno 1984).

Slum upgrading projects in Indonesia and the Philippines were among the largest, often cited by Bank officials as the most successful. The Tondo project in the Philippines, occupying an area of about 180 hectares, was intended to provide for 160,000 slum dwellers.[41] The Kampong project in Jakarta represented one of the largest commitments by the Bank to upgrade slums; at about 1,000 hectares per year, it benefited some 450,000 people annually. By the early 1980s, the project improved infrastructure for more than 60 percent of Kampong's slum areas. More than 200,000 plots were affected, with 3.7 million people benefiting from the project. Costs here were borne not directly by project beneficiaries, but by the local government of Jakarta, which initiated a major effort to increase its fiscal resources through adjustments in property taxation (Ayres 1983; Silas 1984; Baum and Tolbert 1985).

The upgrading project in Lusaka, Zambia, is also cited by the Bank as one of the more successful cases of upgrading in Africa. The program was designed to upgrade four low-income communities (George, Chawama, Chaisa, and Chipta) where about 60 percent of the city's squatters lived. The project cost some US$41.2 million, half of which was funded by the World Bank. Some 17,000 dwellings were improved and 7,600 new serviced sites were provided in adjacent overspill areas, as well as community, health, and educational facilities (Jere 1984; Rakodi 1987). In an evaluation of the project, Bamberger *et al.* (1982) noted that the Lusaka project extended services to almost 20,000 dwellings and provided new services to more than 7,000 families in adjacent spill areas. Zambian authorities were pleased with the project, as it benefited a total of 31,000 families.

Citing these examples, the Bank argued that slum upgrading was economically, politically, and socially less costly than either slum demolition or resettlement. Upgrading also arguably improved the quality of life of the urban poor during the 1970s (World Bank 1991a). However, slum-upgrading schemes did encounter problems of their own. Population relocation was sometimes required in order to widen or pave streets, construct footpaths, or to install water-borne sewerage facilities. Deciding who would be moved and what compensation they should receive was a particular problem. As with sites-and-services projects, tenure and titling remained key issues. The

Bank preferred not to assist in improving facilities and private dwellings unless the occupants possessed the title to the property in question. In Brazil, for example, the Bank opted for sites-and-services over squatter upgrading because of numerous political roadblocks. In Brazilian shanty towns with extremely high population densities, plot demarcation was a prerequisite for successful titling, but this was often precluded by the high squatter population density itself (Ayres 1983; World Bank 1989a).

This chapter aimed to situate the Bank's urban programs of the 1970s within the context of earlier efforts by other international agencies, such as the United Nations and USAID, to address the needs of the developing world's urban poor during the Cold War. The World Bank began to take heed of existing political scenarios and promoted its own version of self-help housing and squatter upgrading schemes. With the World Bank's stamp of approval, these schemes became the new orthodoxy for an entire decade.

While sites-and-services schemes and squatter upgrading from 1972 to 1981 were certainly not panaceas for the Third World's urban housing deficit, they are part of an important moment in the history of the World Bank because they constitute the first and only time the Bank directly targeted the needs of the poor. While the Bank's urban programs did not reach the poorest urban residents, even ardent critics of the World Bank such as Payer (1982) and Caufield (1996) acknowledged that its housing projects did benefit some low-income groups, and discouraged governments from resorting to slum demolition. Seen in this light, the Bank's projects seemed to be practical alternatives to uncontrolled, unserviced settlements, on one hand, and traditional government housing on the other (World Bank 1991a: 27). This was also the first and only time that World Bank programs directly offered the urban poor a framework to legitimize their rights to shelter, infrastructure improvement, and secure land tenure.

The Bank's assessment of its own programs in the early 1980s was generally positive. The programs achieved their objectives at reasonably high rates of economic return, and the Bank recommended continuing its policy. Reflecting on a decade of the Bank's shelter programs, Cohen writes that

> in 1972, given the lack of solutions to urban problems, the strategy appeared sensible as the Bank entered a new sector of lending. In 1982, based on a decade of learning, the strategy is not only sensible: it also offers increasingly promising prospects of success.
>
> (M.A. Cohen 1983: 51)

Other Bank assessments (World Bank 1989a) and external evaluations (Sanyal 1986; Rakodi 1987) of these projects also indicated that, in spite of certain problems, the projects were able to reach poorer groups in all countries concerned.

Nevertheless, for all their success and sensibility, these types of projects began to decline in the 1980s. In the early 1970s, 42 percent of the Bank's urban operations comprised sites-and-services schemes; by the late 1980s, the figure had declined to less than 8 percent. Squatter upgrading followed a similar downward trend; after constituting 30 per cent of the Bank's urban portfolio in the early 1970s, it fell to 3.5 percent in the late 1980s and to just 1.9 percent in the early 1990s (World Bank 1994a). The next chapter provides some possible explanations for this decline by contextualizing it within shifts in the Bank's overall urban agenda.

Chapter 4

The fall of poverty alleviation

The politics of urban lending at the World Bank

For a brief period in the World Bank's history, the urban programs of the 1970s focused primarily on poverty alleviation through direct investment in basic infrastructure and housing for low-income residents. The purpose of these projects was to provide low-cost improvements for the urban poor that could be replicated on a larger scale. According to the Bank's own evaluations, cited in the previous chapter, the programs met their objectives, despite some operational problems. However, in the early 1980s, the Bank began to argue that, although these projects brought certain benefits to the Third World urban poor, they failed to address the complex array of problems confronting the city as a whole. Thus, the Bank began to argue that a shift away from projects and toward a focus on policy was necessary, and that the financial and institutional structures of cities needed to be strengthened instead. Housing assistance, for example, began to move away from shelter projects toward the reform of housing finance policies and the restructuring or dismantling of public housing agencies (Renaud 1983; Richardson 1987b). The Bank began to devote a much larger share of its urban lending portfolio to municipal development projects that sought to "build capacity" and enact financial reforms within municipal governments. By the 1990s, the Bank's urban lending ambitiously took on the challenges of municipal policy, institutional change, and market reform in Third World cities (World Bank 1991b).

The aim of this chapter is to outline some reasons for this change of focus in the Bank's urban agenda. The chapter is divided into four parts. The first part examines how evaluations of low-income housing initiatives helped to shift the Bank's urban policy. The second section outlines bureaucratic changes within the Bank that affected its urban agenda and moved policy away from sites-and-services and squatter upgrading schemes. The third section overviews the conservative political climate of the 1980s and its impact on general policy directions taken by the Bank, such as structural adjustment and the focus on governance. Finally, the chapter discusses how these factors contributed to changes in the Bank's urban agenda. I shall argue that the transformation in the Bank's urban lending program can be attributed to the pressure to keep pace with various trends internal and external to the Bank.

Shifts in World Bank urban policy: internal accounts

In the late 1970s and early 1980s, the urban staff reviewed sites-and-services and squatter upgrading projects on a case-by-case basis, as well as at a more programmatic level (Bamberger *et al.* 1982; Keare and Parris 1982; M.A. Cohen 1983). These reviews were generally positive, as in the following report by Michael Cohen, who coordinated the Urban Poverty Task Force during that period:

> Urban lending, though modest in amount, has had a significant impact on the way urban issues are being analyzed and the solutions formulated and implemented . . .
> Appropriate project design has reduced the cost of providing shelter and infrastructure by as much as 75 per cent and extensive benefits are being generated. Some 1.9 million households have benefited from shelter projects alone . . . The rates of return for these projects are high.

> (M.A. Cohen 1983: 3)

However, as head of the World Bank's Urban Division in the 1990s, Cohen became a strong advocate for a new urban agenda termed "urban management"(discussed later in the chapter). When questioned about the contradiction between his earlier and current views, Cohen claimed that the programs of the 1970s were successful only on a "project by project, case by case basis" but, when analyzed at the city-wide level, they did not have any effect on the multifaceted problems of cities in the developing world:

> The project-by-project approach becomes irrelevant to the scale of the problem. If one projectizes the city, one fundamentally misses the dynamic of scale – so one ends up talking about sites-and-services and not land, you end up talking about neighborhoods and not markets. You don't understand the broader dynamics of the city. The whole notion of moving from projects to urban management was, therefore, essential.[1]

When questioned about the shift, other Bank urban staff also echoed the view that the sites-and-services and squatter upgrading approach was "just a drop in the bucket" of assistance required by Third World cities. They claimed that the projects reached only a small percentage of the target population

– that by the time they were completed, the demand had multiplied fivefold as a result of rapid city growth. Robert Buckley, a housing economist at the Urban Division of the Bank, offered that:

> Today, after more than 20 years of Bank experience and over 100 shelter projects in 60 countries – one can conclude that Bank-supported squatter upgrading and sites-and-services projects have been a considerable improvement over the public shelter programs of the 1950s and 1960s. These projects clearly demonstrated that housing for the poor is possible. However, in spite of their laudable achievements, it is clear that neither sites-and-services nor squatter upgrading can provide the long term answer. Cities are growing too rapidly. What is needed are fundamental changes in policies, institutions, and incentives. This is the only way that resources can be mobilized to meet the demographic demands for housing.[2]

All the urban staff I interviewed in the 1990s echoed Buckley and Cohen, stating that the Bank needed to focus more on systemic features of the city rather than on individual projects per se. Larry Hannah, an economist at the Urban Division since the mid-1970s, explained that "the unit of analysis was primarily the project in the early model. The early projects were developed in very much a responsive mode, in other words, responsiveness to slums, to growth, etc."[3] Hannah regarded the early urban projects as narrow approaches to the problems of cities in the developing world, with the urban staff involved only with particular details pertaining to specific problems. A narrow project focus, according to Hannah, meant that Bank staff mainly dealt with the particular government agency that implemented the project.[4] A number of project staff similarly complained that, while they were working on one particular project, other problems in the cities were being ignored by other departments in that country. Hannah further recalled that

> When we looked at the housing sector in Kenya, we found that policies that had a very important influence on housing were often the responsibilities of ministries with which we had no contact, such as the finance ministry, or the taxation ministry. In many cases, we found that we were dealing with a relatively minor player and had no institutional link to important decision makers.[5]

Thus, there was general consensus among the Bank's urban staff that they needed to focus on institutions and broader management issues rather than individual projects.

While there is some merit in this argument, it curiously makes no attempt to explore how the early poverty reduction programs could have complemented the new urban management focus. Ul Haq[6] (1998) suspects that the Bank had jettisoned poverty-based lending altogether and therefore made no attempt to integrate it into the overall shift in Bank policy in the 1980s. With particular reference to urban lending, McCarney (1987: 149) argues that, in spite of rhetorical statements stressing an interest in assisting the poor, the urban management focus of the 1980s actually signaled the end of targeted lending to the urban poor. In her detailed study of the Dandora upgrading scheme in Kenya, McCarney observes that about 15 percent of the benefits intended for the urban poor "trickled-up to higher-income beneficiaries." Thus, it follows that the non-targeted lending advocated by the urban management staff would be even less likely to benefit the poor. This problem also concerned Edward Jaycox, a predecessor to Cohen as head of the Urban Division.[7] He tried to explain:

> The Bank moved into more systemic issues, such as how to make these cities less centers of consumption and more into centers of production, how to construct a tax base, etc. So policies acquire a greater macro and less micro focus. The more you are involved in systems of cities, the less you are involved with the poor or direct projects.[8]

Despite the Bank's claims, however, the declining significance of sites-and-services schemes and squatter upgrading in the Bank's urban portfolio cannot be explained solely on the basis of internal project evaluations. During the 1980s, as part of their quest to remain relevant in the emerging geo-political scene, the various departments of the Bank restructured their programs to synchronize them with the paradigms of privatization and efficient management adopted by the new Bank president, A.W. Clausen, and his administration. I shall argue that the new urban focus mirrored the Bank's general trend during the 1980s, taking into consideration the bureaucratic reorganization of the Urban Division in response to the conservative political climate of the 1980s.

Bureaucratic reorganization and the fragmentation of the World Bank's urban division

As described in Chapter 3, the Urban Department was first established as a separate unit under the Special Projects Department in 1976. While Bank operations and lending programs were decentralized into country divisions, new units like health, education, and urban projects were managed centrally (see Figure 3.1) in order to facilitate the development and maturation of their agendas into policy. Jaycox recalled that the McNamara years were a vibrant and energetic period in Bank history, especially for the Urban Division:

> We were one of the fastest growing departments ... We could not hire people fast enough because we had so many projects. We were quite a merry band who developed a culture of solidarity with one another.[9]

However, in spite of this seemingly favorable situation, the Bank underwent an administrative reorganization in 1981. After only five brief years as a whole, autonomous unit, the Urban Department was split up into separate "projects" and "policy" sections, with the former falling under the leadership of the vice-president for policy planning and research (Figure 4.1) and the latter being further split across six regional divisions that were physically located in different buildings all over Washington, DC.

Jaycox supported this reorganization at the time:

> We regionalized on purpose. Since the Bank was running into macro-issues in the 1980s, I wanted to ensure that we maintained an urban focus and had some influence at the operational level.[10]

Each regional division received the Urban Department staff in different ways, according to Jaycox. Some regional directors were skeptical of the urban staff and regarded them as idealists who needed a dose of "reality." When pressed about whether regionalization had thrust the urban staff into hostile environments, Jaycox admitted that, although he initially supported the move, he did not anticipate such a reception:

> The kind of work the Urban Division was engaged in did not win us many friends among the traditionalists at the Bank. When we were regionalized, [they] thought they had to control the "crazies" from the Urban Department.[11]

Part of the problem stemmed from the fact that many of the Bank's staff had never regarded urban projects as productive investments all along and were revising their views on project-based lending altogether (discussed in the previous chapter). As Douglas Keare[12] recalls, although the Urban Division tried to reiterate and promote the productive value of housing to skeptical macro-economic operational staff during regionalization, "many people who were not prepared to buy that argument did not buy it."[13] Thus, advocates of urban projects at the Bank found that, in addition to convincing governments of the importance of squatter upgrading, they had to convince the Bank itself yet again.

Meanwhile, the policy section of the former Urban Division was established as a central unit to support and guide the Bank's regional units through their policy work, evaluations of practice, operational support, and basic research.[14] Jaycox left the division in 1981 and was succeeded by Anthony Churchill. This "new" Urban Division served as a kind of internal consultancy unit for the Bank, according to Cohen, a research arm that "identifies directions for lending in the country division."[15] Keare, however, argues that "this whole conception of the urban division as 'new' is somewhat self-serving" because it was not able to initiate new programs as it did in the 1970s, or exercise the clout it enjoyed during the McNamara years.[16]

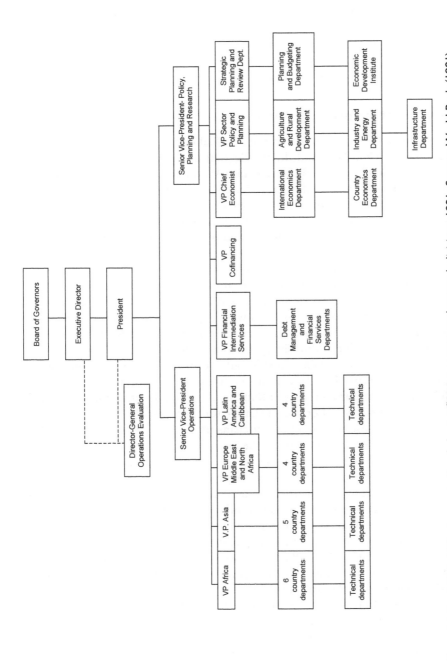

Figure 4.1 Partial World Bank organizational chart reflecting country and research divisions, 1981. Source: World Bank (1981).

Seen in light of these changes, it becomes apparent that sites-and-services and squatter upgrading schemes suffered twin internal setbacks as major urban projects: the problem of regionalization in 1981 was caused by its coincidence with the Bank's reconsideration of the priorities of the 1970s. However, a third factor, the emerging conservative political backlash against the Bank's operations, did not augur well for sites-and-services or squatter upgrading programs as policy options for housing the Third World urban poor. This issue is considered below.

Conservative policy reform

It was argued in Chapter 2 that the political climate of the 1960s – decolonization, the Cold War, and the intellectual and political recognition of underdevelopment – prodded the reluctant World Bank toward social lending. However, the elections of Margaret Thatcher in 1979, Ronald Reagan in 1980, and Helmut Kohl in 1982 heralded a new ideology that would soon seize the World Bank and urban policy. Thus, the urban programs of the Bank during the 1980s ought to be seen in the context of the political priorities and pressures of that decade.

McNamara's poverty alleviation initiatives were not without controversy. His own relationship with the Nixon administration was a tenuous one; Nixon half-heartedly supported McNamara's bid for a second term as president of the Bank, arguing that he was not always supportive of American interests. Clark notes that McNamara "had irritated many important leaders in the Nixon administration," who "thought that as an American, McNamara should be more responsive to policy nudges, but over time he proved unnudgeable" (Clark 1981: 176). In 1976, Treasury Secretary William Simon opposed McNamara's call for increases in Bank lending and disagreed in public with McNamara about whether such a change was necessary.[17]

There were also a number of criticisms in the late 1970s and early 1980s from other conservative quarters – the financial press and conservative think-tanks – against the Bank's anti-poverty programs. The conservative Heritage Foundation, for example, in its study Agenda for Progress, urged the United States to withdraw its support from multilateral development agencies like the World Bank and focus instead on bilateral aid programs. This view was shared by Peter Bauer,[18] who complained that the Bank's poverty programs focused more on "redistribution and restitution" than on sound economic policy (Bauer 1979: 462). Bauer accused the Bank of encouraging state intervention at the expense of the private sector in developing countries and argued that "bilateral grants are less likely to become instruments of worldwide egalitarianism." According to Bauer, the United States would best be served by bilateral programs over which it could exercise greater control. These, in turn, would better reward market-oriented economic policies in recipient countries. Schmidt (1979) felt that multilateral development agencies tended to be malleable under political pressure and engaged in economically irrational projects because they were too insulated from the pressures of the market. As a remedy, Schmidt recommended privatizing multilateral development banks in order to depoliticize them and enforce market discipline upon their operations.

In May 1980, *Forbes* magazine launched an attack against the Bank, arguing that its "so-called new-style projects" were not pro-growth but rather welfare projects for the developing world which undermined the Bank's ability to impose fiscal discipline on the economic policies of member countries. The article quoted an anonymous World Bank official who compared developing nations to drunks: "You don't persuade him to drink less just by giving him more money."[19] In June 1980, *Barron's* joined in, charging that under Robert McNamara's direction the Bank was more concerned with the quantity than with the quality of individual projects and was, therefore, unable to effectively influence market-oriented changes in recipient countries.[20] Echoing that view in a publication of the Heritage Foundation, Phaup (1984: 13) stated that "it appeared as if McNamara thought that an 'International Great Society' could be established if only enough money and good intentions were devoted to the effort." Phaup denounced the Bank's anti-poverty programs as "unproductive, socialistically inclined experiments" that amounted to "dependence-inducing international welfarism" (ibid.). Thus, conservatives railed against the World Bank for emphasizing too many "basic needs" projects such as agriculture, education, and urban shelter. In their view, these areas forced the Bank to rely heavily upon government intervention in order to resolve economic problems. By emphasizing social concerns, the Bank was allegedly encouraging and supporting imprudent policies that discouraged private investment and buttressed the forces of statism. In sum, McNamara's opponents charged that he had turned the World Bank into a soup kitchen that dished out doles to the developing world.

Conservative criticisms and changes in the World Bank's leadership

In June 1980, McNamara announced, rather abruptly, that he would be leaving the Bank in the following year, before completing his third five-year contract. While McNamara cited personal reasons for his departure, Bank historian Jochen Kraske notes that political considerations also played a role in McNamara's decision: McNamara felt that he could not provide continuous leadership at the Bank in light of the conservative climate of the period and retired as president of the Bank in June 1981.[21] The Bank's incoming president, A.W. Clausen, was chosen by President Jimmy Carter in consultation with the Reagan presidential transition team (Rowen 1994). Clausen's appointment, therefore, coincided with this period of intense conservative criticism against the Bank, especially within the USA, the Bank's largest donor.

When he assumed the American presidency in January 1981, Ronald

Reagan and his conservative senior advisors gave legitimacy and prestige to the mounting ideological and political assault on multilateral development institutions, particularly the World Bank. Previous US administrations, both Democratic and Republican, had mostly supported the Bank as an important instrument of US foreign policy in spite of disputes over specific issues. In this regard, the Reagan administration's advent represents a "breakpoint in US policy toward the Bank" (Gwin 1994: 37). After Reagan's electoral victory in 1980, the new administration signaled its intention to emphasize bilateral over multilateral development aid programs by actively seeking a reduction in US commitments to multilateral development agencies. In 1981, Reagan appointed Edward J. Feulner, Jr., President of the Heritage Foundation, as his advisor on international development assistance. Feulner promptly called for a de-emphasis of the American role in multilateral institutions in general and targeted the Bank's projects in particular as "social experiments" whose "utopianism . . . American taxpayers should not be asked, directly or indirectly, to subsidize."[22]

Reagan's first director of the Office of Management and Budget, David Stockman, wrote in his memoirs that

> The organs of international aid and so-called Third World Development – the UN, the multilateral banks, and the US Agency for International Development – were infested with socialist error. The international aid bureaucracy was turning third world countries into quagmires of self-imposed inefficiency.
>
> (Stockman 1986: 116)

During the first week of the Reagan presidency, Stockman called for the elimination or reduction of American participation in multilateral development agencies "that are not responsive to US foreign policy concerns" and are "ineffective in producing sound economic development" (ibid.: 57). Stockman's first proposal was to cut US bilateral and multilateral aid by 45 percent, a US$13 billion reduction for the 1982–6 period. He also wanted to renegotiate Jimmy Carter's commitment to the IDA and proposed to terminate all US contributions to the soft loan windows of the IDA and other multilateral development banks. However, State Department pressure and political compromises between the various arms of government eventually diluted Stockman's proposals (Stockman 1986; Gwin 1994).[23]

United States Treasury Undersecretary Beryl Sprinkel also argued against governments or international developmental institutions taking on functions that could be performed by the private sector (Rowen 1994: 300). One of her first acts in office was to commission a Treasury Department study on the policies and operations of multilateral development institutions, including the World Bank. The purpose of the report was to ascertain, among other things, whether the private sector was being crowded out by government-to-government loans, and to determine if the World Bank had socialist tendencies. While the final report was generally supportive of the Bank, it did mention that "the Reagan Administration brought a fundamentally different view of both the need for and underlying economic philosophy of all government programs" and that US support for and participation in organizations like the World Bank "should reflect its economic philosophy as to how economic development can be effectively promoted" (Department of Treasury 1982: i).

While the Treasury Department did recognize the importance of multilateral agencies like the World Bank in promoting economic growth and stability in the developing world, it stressed that American support for such organizations ought to be contingent upon whether their programs and operations "encourage adherence to free and open markets [and] emphasi[ze] the private sector as a vehicle for growth and minimal government involvement" (ibid.: 7). The report made three key recommendations. First, the United States should continue to support multilateral development agencies only if their policies did not inhibit private sector development. Second, loan allocations should be conditional on policy reforms in developing countries and urged multilateral agencies to "exercise their leverage more effectively" (ibid.: 80). Greater cooperation between the Bank and the IMF was stressed in order to achieve improved economic performance in less developed countries, especially in promoting structural adjustment to secure major policy reform. Third, the United States ought to reduce its commitment to multilateral agencies altogether in the long run. The report recommended eight specific proposals, which are summarized in Table 4.1.

Looking back at the McNamara years, the Reagan administration and its allies in Congress objected to the tremendous growth of the Bank and the IDA during this period. Stockman (1986) argued that the Bank emphasized income redistribution and supported communist regimes under McNamara, while paying insufficient attention to the security concerns of the United States. In his opening address at the annual meeting of the World Bank's board of governors in 1983, President Reagan agreed to continue supporting commitments made to the Bank by the Carter administration but stressed that the United States would oppose Bank programs that supported either government-led public sector development or socially oriented development projects. "What unites successful countries is their belief in the magic of the market place," Reagan declared. "Millions of individuals making their own decisions in the market place will always allocate resources better than any planning process."[24]

The Bank's leadership was thus confronted by a very different political climate in the 1980s than that which McNamara had faced in the 1970s. During the 1970s, the interrogation of traditional models of development, the political concern that poverty would facilitate the spread of communism, and international forums like the Pearson and Brandt Commissions all pushed the

Table 4.1 US Treasury Department recommendation for policy changes in programs and operations of multilateral development agencies

Market forces	Lending policies and programs should increasingly focus attention on market signals and incentives, on private sector development, and on greater financial participation by banks, private investors, and other sources of private financing (with particular emphasis on the International Finance Corporation's approach and type of program)
Promoting policy reform	Annual country and sector lending levels should be more flexible and less target oriented. The banks should be encouraged to introduce more selectivity and policy conditionality within the projects and sector programs they support. The banks should concentrate their lending in those sectors where they have the most expertise and where they will have the most policy leverage. The IBRD structural adjustment lending program and the IDB sectoral lending program should be closely monitored to assess the potential for achieving policy reform
Sector allocation	The multilateral development bank (MDB) sector allocation process should be based upon the economic or social priorities of the borrower government – but only to the extent that their priorities are consistent with the basic economic principles of the MDBs. Increased emphasis should be given to the economic rate of return and policy improvements that can be obtained in the project selection process, and lending should increasingly be linked to specific policy reforms. Due consideration should also be given to the relation of the project to overall development strategy, and to its overall economic impact
Graduation	Existing IBRD graduation policy should be implemented more effectively and emulated in the regional banks. Countries above the agreed income threshold should be phased out over a reasonable period, with the Bank assisting the graduation candidate to remove the remaining constraints to self-sustaining growth during the phase-out period

Maturation	The USA should encourage better allocation of soft lending funds through a more systematic "maturation" policy. This would require obtaining agreement to move countries into the hard loan windows as rapidly as their debt servicing capacity permits, also giving due consideration to the impact on the loan portfolios of the banks' hard loan windows
Paid-in capital	The USA should develop and implement a plan to phase down and eventually phase out new paid-in capital for the hard loan windows on a schedule consistent with maintaining the financial integrity of the MDBs
Soft loan window	The USA should begin to reduce its participation in real terms in the soft loan windows, especially the IDA because of its large share of the budget, but at a pace consistent with US economic and political objectives. This is consistent with our desire to realign the concessional windows more closely to assisting the poorest developing countries, and saves budgetary resources. We should also place increased emphasis on the adoption of effective policies by the remaining soft loan recipients
US budget priorities	The following factors should be carefully considered in determining US allocations to the institutions: • To the extent needed, callable capital allocations to hard loan windows are preferable to concessional window replenishments. • The World Bank group has been more successful than the regional banks in promoting appropriate economic policies. • Among the regional banks, the Inter-American Bank ranks particularly high in terms of our political/ strategic interests. The Asian Bank ranks highest in institutional efficiency. The African Bank ranks high in terms of humanitarian concerns

Source: Department of Treasury (1982: 7–8).

international development community to address socioeconomic inequality. These factors collectively convinced the McNamara administration at the Bank that addressing poverty, especially in urban areas, was vital. Sites-and-services and squatter upgrading programs became part of the Bank's urban agenda under McNamara because of these considerations. In contrast, the political atmosphere of the 1980s was not very sympathetic to development assistance in general and poverty reduction in particular. The decidedly rightward shift in the international political scene with conservative electoral victories in the the United States and several European countries meant that the World Bank and its new president, A.W. Clausen, had to convince skeptics in Washington of its integrity and relevance.

In 1980, Munir Benjenk was appointed vice-president of external relations at the Bank following Clausen's appointment. Calling himself "a closet conservative of the McNamara era," Benjenk approached the new president to discuss an appropriate image for the Bank. He advised Clausen that, in light of the new political environment, it would be strategic for the World Bank to avoid being identified with the McNamara era. Clausen was thus advised to signal a break with the past and "make peace with the conservative public opinion" (cited in Kapur *et al.* 1997: 336). This idea resonated well with Clausen's own conservative inclinations. Ul Haq recalls his dismay during a meeting with Clausen in which the new Bank president expressly stated that poverty-directed lending would not figure prominently during his administration, and that the poverty focus would be "no more than a thin veneer."[25]

In his first annual address to the board of governors, Clausen stressed that developing countries ought to "structural[ly] adjust to the realities of the global economy," and that the Bank should "direct its project and sector lending to assist developing countries in making these structural adjustments" (Clausen 1986: 8–9). He also confirmed that the Bank would encourage private sector initiatives. In January 1982, the Brookings Institution sponsored a conference on how the Bank might adapt to the political and economic environment of the 1980s, especially the skepticism from the USA. Clausen went to great lengths in his address to stress the virtues of the market and his admiration for the private sector:

> I think you all know pretty well where I stand. After all, the private sector is what I know best and what I have called home for more than thirty-one years. As a commercial banker, my whole career was spent in that competitive, creative, energetic marketplace. I have to say honestly that I loved it, and still do . . . When anyone says that the private sector ought to be more involved in the development effort, I am always the first one in the congregation to say "Amen." I know it works.
>
> (Clausen 1982a: 67)

It is ironic that Clausen, a fierce advocate of private sector-driven development, had personally relied on public sector assistance in the form of grants from the GI bill for his own education (Kraske 1996).

Clausen also questioned the "old" north–south paradigm that undergirded some of the Bank's thinking in the late 1960s and 1970s. The Pearson (see Chapter 2) and the Brandt Commissions had stressed the need for the international community to address north–south disparities and had influenced the Bank's development thinking in turn. However, as part of the paradigm shift of the 1980s away from international development assistance, the north–south model was abandoned in favor of fiscal responsibility, governance, and market-based reforms (N. Harris 1986; Raffer 1999). In a speech to the Yomiuri International Economic Society in Japan, Clausen proclaimed that

> the old North–South economic model of the international economy of the 1960s and the 1970s is no longer very useful. It is not very useful because it has tended to create a bipolar concept of world economic dynamics that glosses over – or completely leaves out – a whole series of other elements of economic activity that just do not fit into a rigid North–South dichotomy.
>
> (Clausen 1982b: 1–2)

Citing the "success" of the newly industrializing countries (NICs), Clausen stressed that the Bank should pursue innovative policy agendas designed for the realities of the 1980s. His emphasis on stabilization, balance of payments, privatization, fiscal discipline, and sound productive projects echoed back to Eugene Black's vision for the Bank (discussed in Chapter 2).

> The World Bank, of course, is – and will remain – a bank. A very sound and prudent bank ... It is not in the business of redistributing wealth from one set of countries to another set of countries. It is not the Robin Hood of the international financial set, nor a giant global welfare agency. The World Bank is a hardheaded, unsentimental institution that takes a very pragmatic and nonpolitical view of what it is trying to do.
>
> (ibid.: 14)

Thus, Clausen effectively steered the Bank away from McNamara's agenda, stressing free markets during his tenure as president. In doing so, he made a strong neo-liberal impression on the Bank's economics, its staff, and its operations.

Changes in key senior appointments at the Bank reflected the contemporary political tenor and helped to shape the Bank's new priorities for development. Hollis Chenery, vice-president for research and senior economist under McNamara, left soon after Clausen was appointed president. McNamara and Chenery had hoped to settle Chenery's replacement prior to McNamara's departure with the nomination of Albert Fishlow, a Yale University economist, for the position. According to World Bank historian Jochen Kraske, Clausen initially endorsed the appointment but withdrew his support after the *Wall Street Journal* launched a vicious attack on Fishlow.[26] The *Journal*'s editorial lashed out that "McNamara [was] trying to pack the world's most important lending institution" with his "ideological clones . . . in an effort to jam the bank on automatic pilot before the next guy takes over." Fishlow was portrayed as a Leftist ideologue whose intellectual work focused on politically taboo ideas like "income distribution in Brazil." The editorial complained that "US taxpayers who support the World Bank have seen too few success stories among the nations that have taken the largest doses of McNamara's advice and money" and recommended that the Bank's next vice-president for research ought to be someone who moved the Bank away from McNamara's agenda.

In response to this controversy, Clausen appointed Anne Krueger, an economist from the University of Minnesota, to the position.[27] Krueger immediately reshuffled the staff of economists at the Bank and emerged as a powerful spokesperson for "new" policy priorities, among which issues of equity and poverty were absent. As a neo-classical economist who strongly emphasized pro-market, anti-interventionist development priorities, Krueger was not without controversy herself. Krueger went to great lengths to impress her views upon the economics research unit and, in the process, stifled the open intellectual inquiry necessary for sound research and frustrated staff with her intolerance for dissent (Kapur *et al.* 1997).

Finally, Mahbub ul Haq resigned as director of policy planning in 1982 on the grounds of irreconcilable differences with the Bank's new president. Ul Haq's departure was significant because he was one of the Bank's most forceful advocates for poverty-oriented lending during the McNamara years. Hired by McNamara in 1970 (as discussed in Chapter 2), ul Haq became a part of McNamara's inner circle and was noted for drafting McNamara's influential Nairobi address.[28] Ul Haq ultimately left in frustration because he believed that he could not affect policy while the Bank was subservient to hardliners in the American administration in Washington. "The goddess of growth is back on the pedestal," he told the *Sunday Times*. "Right wing economics is destroying all that the Bank achieved in the McNamara years."[29] Unimpressed by Reagan's "magic of the market place" formula, ul Haq was deeply concerned that the Bank's overemphasis on the role of private investment and market-driven development would condemn the poor to "absolute, intolerable poverty, because they are at best on the fringe of the market-place, more often outside it altogether." He saw the Bank's post-McNamara policies as a betrayal of the Bank's very reason for existence, . . . It is after all the Bank for Reconstruction and Development. If it sits back and hands over to the market, it will abdicate its responsibilities – and its duty is to provide a financial cushion for reform.[30]

Thus, social equity concerns in the Bank diminished with the departure of major proponents like ul Haq. The new leadership's strong emphasis on private sector-driven development, free markets, and anti-statist programs resulted in an agenda that increasingly embraced the now infamous structural adjustment policies toward the developing world.

Structural adjustment, governance, and the private sector

While there was growing concern in certain quarters of the Bank that specific project-based lending did not influence broader policy in the developing world, it was not a totally novel issue at the Bank. In fact, in 1965, during the Wood Presidency, a Bank mission to India recommended major policy reform, including devaluation of the rupee, relaxation of administrative controls, and the promotion of the private sector (World Bank 1965). Furthermore, the Bank never made unconditional loans; project loans always carried stipulations. For example, the Bank's support of a power station project might be contingent on certain adjustments to electricity tariffs. However, in the 1980s, the Bank launched a new form of development financing called structural adjustments loans (SALs). SALs expanded the scope of conditionality in lending, whereas most of the conditionalities before the 1980s were only sectoral or subsectoral in scope. SALs were different in three important ways. First, the SALs were non-project loans, which meant that finance was separated from the specific items of investment. Second, program lending was combined with specific policy change conditions. Finally, these conditions were not just sectoral and subsectoral, but national and macroeconomic in scale and nature. In other words, these loans were contingent upon major economic reforms in a particular developing country (Mosley *et al.* 1991). SALs are considered in some detail below.

Structural adjustment lending became prominent at the Bank when Clausen replaced McNamara and the views of the new chief economist, Anne Krueger, began to influence Bank policy. While the idea itself was introduced by Robert McNamara, the structural adjustment policy deviated in practice from what McNamara and others had originally conceived and became increasingly affiliated with Clausen's agenda for the Bank (Kapur *et al.* 1997). A publication from Krueger's unit stated that "tolerance for slipshod policies during the romantic 1970s are part of the history of development; it behooves the Bank to inject rationality into this process" (World Bank 1984a: 12).

Babai (1984: 280) observes that the World Bank used the term "structural adjustment" in two ways. On one hand, it denoted an approach to economic growth "that involves fundamental change rather than marginal tinkering," was comprehensive in scope, and sought to adjust the underlying structures of the economy. On the other hand, the term refers to adjustments that are responses to structural problems in the world economy, such as oil price hikes and the debt crisis. SALs then, in the Bank's view, were structural remedies to structurally induced crises. The process of adjustment itself involved two stages, according to the Bank (World Bank 1989b: 2). First, a country had to acknowledge the reality of the challenges facing it, a reality that the country had previously denied, according to the Bank. Second was the need to reshape and redirect policies and institutions in LDCs. Adjustment begins, according to the Bank, when old notions of development are abandoned in favor of new, growth-oriented reform.

In light of this "new" view at the Bank, there was growing sentiment that the "old" project-based approach was not adequately injecting the "rationality" of macroeconomic policy. In 1978 itself, Ernest Stern, the chief of operations at the Bank, had recommended upon advice from Stanley Please, his senior advisor on structural adjustment, that certain country projects ought to be reduced or delayed because of their lack of attention to macro-economic policies.[31] Please contrasted development lending with SALs as follows:

> Development programs embody measures – improved provision of infrastructure, technological change, education, health, population, and so on – that are required to ease the basic constraints on growth and development and, therefore have a long-term focus. Adjustment programs on the other hand . . . aim at achieving viability in the medium term . . . [They] ask the question – how can the existing productive capacity be used more efficiently?
>
> (Please 1984: 18)

For Please, SALs were necessary not only because they addressed short-term balance-of-payment issues, but because long-term goals might be realized through adjustment lending. SALs, then, were intended to fundamentally reform economic policy in four ways: first, by improving a country's economic climate and its capacity to attract foreign investment by eliminating trade and investment barriers; second, by reducing government deficits through spending cuts; third, by boosting foreign exchange earnings by promoting exports; and, fourth, by ensuring that debt is serviced (George and Sabelli 1994).

However, according to Please, the Bank was always confronted by the challenge of linking SALs with project-related activities, which were still deemed necessary for repairing practical, short-term problems: "Does project lending provide an effective means through which the Bank can help governments formulate programs of structural adjustment and monitor their implementation?" (Please 1984: 25). A review of the record indicated otherwise to Please because the "tail of project operations has been able to wag the dog of policy and of broad institutional reform" (ibid.: 26). Thus, Please was part of an increasingly vocal group during the later years of the McNamara presidency who fought to bring domestic policy reform to center-stage at the Bank, and pushed for a policy dialogue with the developing world that went beyond gentle persuasion (Mosley *et al.* 1991; Mosley and Eeckhout 2000).

McNamara's 1979 speech at the United Nations conference on trade and development in Manila represented the first time that senior policy-makers at the Bank connected project loans directly to major policy reforms. In fact, it was during this speech that the term "structural adjustment" was first publicly proposed by the Bank as a means of influencing policy in the developing world:

> In order to benefit fully from an improved trade environment, the developing countries will need to carry out *structural adjustments* favoring their export sectors. This will require appropriate domestic policies and adequate external help.
>   I would urge that the international community consider sympathetically the possibility of external assistance to developing countries that undertake the needed *structural adjustments* for export promotion in line with their long term comparative advantage.

> (McNamara 1981: 549, emphasis added)

When I asked him about structural adjustment in his speech, McNamara said that it was his intention to stress two interrelated issues. First, he wanted to emphasize that developed countries ought to do more to open their markets to manufactured exports from the developing world. Second, he tried to encourage the developing world to reform policies that inhibited effective participation in the international marketplace. McNamara insisted that while he recognized the need for structural adjustment, he did not intend it to override the Bank's focus on poverty or adversely impact the Third World poor.[32]

It is interesting to note here how McNamara's use of the term "structural adjustment" differs from later usages. Bank officials after McNamara have frequently deflected criticism of structural adjustment by arguing that it was initiated within the Bank by McNamara himself and not by external political pressures. By doing so, however, they refuse to acknowledge how their definition and implementation of structural adjustment deviates from McNamara's original conceptualization: McNamara maintained his focus on poverty and emphasized the actual opening up of markets in the developed world to Third World imports, whereas since Clausen structural adjustment has meant a revival of older schools of thought that stress comparative advantage, ignore poverty, seldom mention the role of developed economies, and never discuss the diversification of developing economies. In an interesting semantic twist, the World Bank under Clausen appropriated its own term, "structural adjustment," from McNamara and restored older, more conservative meanings.

After McNamara, SALs constituted about 30 percent of the Bank's lending in the 1980s, and about 60 percent during the 1990s, rising from less than 10 percent in the early 1980s.[33] This increase has made SALs the most prominent and contentious part of the Bank's lending programs. (Table 4.2 provides an overview of adjustment lending grouped by country.)

In addition to these issues at the Bank, a number of external factors such as the oil and debt crises had become important. The IMF moved to the forefront of international finance as it played a leading role in providing loans in exchange for policy reform in developing countries. As these events pushed the World Bank to define a central role for itself, SALs seemed to provide the perfect entry point into the financial affairs of member nations while conforming to the policy recommendations of the Reagan administration (Feinberg 1988).

The transition toward SALs in the Bank during the late 1970s and early 1980s is clearly reflected in the Bank's report on Africa's economic crisis, *Accelerated Development in Sub-Saharan Africa* (World Bank 1981). The "Berg Report,"[34] as it is frequently called, shaped the thinking of multilateral development agencies toward Africa during the 1980s and strongly reinforced the rationale for structural adjustment lending. While the report cited a few domestic structural barriers to development in Africa, such as underdeveloped human resources, hostile climates, and high population growth rates, as well as a few external impediments such as high energy prices and low demand for African primary commodities, the major obstacles to African development identified by the report were "domestic policy deficiencies" and "administrative constraints" (ibid.: 41). The report attributed Africa's economic crises mainly to domestic policy choices, which provided inadequate incentive for agricultural growth. Furthermore, the report argued that national policies discouraged the private sector from making a contribution to national development, thereby placing an overwhelming burden on the public sector. The lack of administrative and managerial talent in the public sector was also seen as compounding the problem, according to the report. As a consequence, the report concluded, Africa's market efficiency was slowed down. The "Berg Report" recommended a wide range of policy reforms to facilitate market efficiency by stimulating agricultural outputs and exports, with the argument that agricultural growth in predominantly agrarian societies would accelerate overall economic growth because agriculture was their comparative advantage. The report added that the size and economic

Table 4.2 World Bank structural adjustment lending

Loan and borrower	1980–2	1983–6	1987–90	1991–3
Adjustment loans (AL)	1,412	3,553	5,597	4,744
Adjustment loans/total loans (%)	7	18	26	23
Structural adjustment loans/total AL (%)	87	40	45	51
Sectoral adjustment loans/ total AL (%)	13	60	55	49
Borrowers				
Africa	320	916	1,305	1,049
Percentage of total AL	23	26	23	22
No. of loans	3	10	18	14
East Asia	301	389	687	147
Percentage of total AL	21	11	12	3
No. of loans	1	1	3	924
Europe and Central Asia	440	572	498	19
Percentage of total AL	31	16	9	4
No. of loans	1	2	2	1,527
Latin America and Caribbean	95	1,257	2,284	32
Percentage of total AL	7	35	41	10
No. of loans	2	5	9	474
MENA	0	229	437	10
Percentage of total AL	0	6	8	2
No. of loans	0	1	2	621
South Asia	256	189	386	13
Percentage of total AL	18	5	7	4
No. of loans	2	1	3	1,743
Highly indebted countries[a]	165	2,020	3,015	37
Percentage of total AL	12	57	54	13
No. of loans	1	7	11	13

Source: Kapur et al. (1997: 520).

Note
a Highly indebted countries are Argentina, Bolivia, Brazil, Chile, Colombia, Costa Rica, Cote d'Ivoire, Ecuador, Jamaica, Mexico, Morocco, Nigeria, Peru, Philippines, Uruguay, Venezuela, and Yugoslavia (until April 1993). Does not include 1993 loan to Slovenia for US$80 million.

responsibilities of African states ought to be curtailed if the reforms were to be effective.

The assessments of the "Berg Report" were in direct conflict with a report released by the Organization of African Unity (OAU) a year earlier. The OAU report found that Africa's low position in the international economic order was largely due to negative external conditions. As these two reports held contrasting views on the root causes of Africa's economic crisis, there were a number of rhetorical battles over their conclusions. However, the "Berg Report" was significant because, in effect, it had broken a taboo: "Never before had the Bank [been] publicly critical of such a large group of borrowers" (Kapur *et al.* 1997: 719). Since Africa's domestic policies were allegedly at fault, the report's claim was that SALs were on target as a policy measure. In spite of the torrent of criticisms, the "Berg Report" led Bank policy on Africa during the 1980s and provided a rationale for structural adjustments in the continent. Lancaster (1987) notes that, in spite of African reservations and objections to it, a number of Western governments welcomed the report because it corroborated their own analyses of Africa's economic problems, resonated with their ideological positions, and conformed to their existing economic relationships with Africa.

In sum, the Clausen administration succeeded not only in downplaying poverty reduction as the Bank's mission, but also in changing the tone and substance of the Bank's message during the 1980s to reflect economic liberalization in LDCs. As part of that shift, Clausen's tenure at the Bank was characterized by the expansion of policy-based lending, with structural adjustment as its main focus.

In September 1987, Barber Conable replaced Clausen as president of the Bank.[35] While maintaining Clausen's focus on markets and fiscal restructuring, Conable tried to respond to mounting criticism of structural adjustment lending and steer the Bank away from its new image as the "world's debt policeman."[36] In his first address to the board of governors in 1987, Conable mentioned that the focus on alleviating poverty "would be reaffirmed, reintegrated and re-vitalized." He spoke of a "balanced development program" at the Bank and was concerned about its fixation with adjustment.[37] Anne Krueger resigned as vice-president of the Economics and Research Unit in the fall of 1987. While her replacement, Stanley Fischer, was neither passionate about poverty reduction like ul Haq nor an ardent supporter of Chenery's basic needs approach, he did elevate the priority of poverty reduction in the Bank's research agenda, as reflected in the Bank's 1990 world development report on poverty (Kraske 1996; Kapur *et al.* 1997).

After citing statistics on global socioeconomic inequality, the *World Development Report (Poverty)* (World Bank 1990: iii) recommended a two-pronged strategy to attack poverty. One was the pursuit of economic growth that ensured "productive use of the poor's most abundant asset – labor." The other was the provision of basic social services such as primary education and health care to the poor. In this report, the Bank under Conable tried to reconcile the Clausen years' emphasis on growth and trickle-down economics with some elements from the targeted poverty-oriented programs policies of the McNamara period.

However, from the very start, the Conable administration found it difficult to reconcile a focus on poverty with its operational commitments; in fact, poverty lending encountered the same resistance it had faced during the later McNamara years and the Clausen presidency (Caufield 1996; de Vries 1996). Given the prevailing conservative political atmosphere and the discrediting of McNamara's poverty reduction efforts, poverty concerns sat uneasily with adjustment, growth, and free markets because of a number of constraints. First, the Bank's senior managers from the Clausen years were conservative and maintained their skepticism toward poverty-oriented approaches. Second, Conable, under pressure from the US administration, subjected the Bank to a traumatic reorganization of staff that soon impeded his own leadership. Two years after the reorganization, Conable was replaced by the former chairman of Morgan Guarantee, Lewis T. Preston, who was a lifelong banker like Clausen.

While Preston initially considered whether the Bank should be concerned with poverty reduction as an "overarching" objective, he was not enthusiastic about committing the Bank to a broad social agenda (Kapur et al. 1997). There was also a discrepancy between the senior management staff's public rhetoric and internal statements on poverty lending. For example, Ernest Stern, now Managing Director of the Bank and Chairman of the Key Operations Committee under Preston, publicly expressed optimism about the Bank's poverty reduction focus but privately brushed it aside. In an address celebrating the fortieth anniversary of the Bretton Woods institutions, Stern remarked:

> How the benefits of growth are redistributed, is, in fact, a central issue of development. Central, not only because it affects long-term prospects for political stability and national cohesion, but central because it is the major reason why the development community is concerned about development.[38]

However, Stern wrote in an internal memorandum that "poverty alleviation is but a minor objective most of the time," which he reiterated in his private farewell address to the Bank:

> I do not think we should ever confuse development assistance with charity because charity cannot be a concept among nations ... that is why I believe that a Bank, requiring repayment, anchored in financial discipline, supported by sound project analysis, by the monitoring and evaluation of results and by a sense of accountability, is the best channel for assisting development effectively.[39]

The Bank's renewed emphasis on poverty reduction came at a time when the institution was under attack for exacerbating socioeconomic inequality in the developing world. For example, resounding objection to the Bank's structural adjustment programs had come from the United Nations Children's Fund (UNICEF), which published a collection of essays in 1987 called *Adjustment with a Human Face*. The work chronicled the social costs of adjustment and the worsening conditions of health, education, employment, and incomes in countries participating in the Bank's adjustment programs. As the Bank was preparing for its fiftieth anniversary celebrations, an NGO coalition called Fifty Years is Enough posted full-page advertisements in major newspapers criticizing the Bank's policies. Miller-Adams (1999: 87) found that the criticisms were not well received at the Bank, whose staff members felt that they were genuinely motivated to help countries, and that they emphasized economic rationality because "that was the only model they understood."

Such an atmosphere of critique caused the Bank to tone down its emphasis on economic adjustments per se. Nelson (1995) notes that some policy changes were made, including an emphasis on gender and environmental issues as well as greater collaboration with NGOs. However, the core policy concerns of the Bank during this period, such as the private sector and governance, remained well within the bounds of the neo-liberal tradition.

The issue of "good governance" began to dominate development thinking in the mid to late 1980s, with a flood of academic texts and conferences on the subject (Jackson 1977; Carter Center 1990; Hyden 1992; Hyden and Bratton 1992). However, the World Bank's articles of agreement prevented the Bank from addressing political issues directly. The founders of the Bank had gone to great lengths to emphasize the neutrality of the institution with regard to political ideologies and interests:

> The Bank and its officers shall not interfere in the political affairs of any member country, nor shall they be influenced in their decisions by the political character of the members concerned.[40]

Nevertheless, as "governance" acquired a prominent position in the development lexicon of the 1980s, the World Bank also adopted the term into its own agenda. If the Berg Report, discussed earlier, marked a watershed with the Bank's venture into structural adjustment lending, another report on Africa, *From Crisis to Sustainable Development* (World Bank 1989b), heralded the emergence of governance as a key issue for the Bank in the late 1980s. While this report concurred with the need for appropriate economic policies, it went further:

> Underlying the litany of Africa's development problems is the crisis of governance. By governance is meant the exercise of political power

to manage a nation's affairs. Because countervailing power has been lacking, state officials in many countries have served their own interests without the fear of being called into account. In self-defense individuals have built up personal networks of influence rather than hold the all-powerful state accountable for its systemic failures.

(World Bank 1989b: 60)

Africa needed a "pluralistic institutional structure" for economic and political renewal, the report argued, because "history suggests [that] political legitimacy and consensus are a precondition for sustainable development" (ibid.: 61). The report also identified civil society as an important dimension of governance:

[Civil society has] an important role to play . . . [it] can create links both upward and downward in society and voice local concerns more effectively . . . [it] can also exert pressure on public officials for better performance and greater accountability.

(ibid.)

The World Bank saw accountability, legitimacy, transparency, and participation as a means of forging new scaled linkages and reducing the power of the national state. To this end, the Bank encouraged decentralized administration and the strengthening of local government. These themes became integral to the Bank's new urban agenda and are discussed below.

The concern with governance evolved, in part, in response to the problems the Bank experienced with structural adjustment lending. The Bank stated that the success of SALs were limited by political issues that exceeded the technical capacity of the public sector to handle them. According to the Bank, SALs widened the scope of its lending to encompass the realm of politics, as "political factors have a decisive influence on the choice of policies dealing with macroeconomic disequilibrium" (World Bank 1990: 115). The Bank's interest in governance is linked to private sector-driven strategies of the SALs – both of which are intended to promote free and unimpeded markets in the developing world (Miller-Adams 1999).

The twin concerns of SALs and governance were reinforced by the collapse of existing socialism in Eastern Europe and the Soviet Union. After 1992, a new group of countries became members of the Bank and began to receive considerable attention. In addition to competing for resources with the developing world, these new member countries also began to affect the nature of the Bank's operations. After the demise of reconstruction lending in the mid-1950s, the Bank was primarily concerned with lending as a means of staving off communism in the developing world. After the Cold War, Eastern European and Central Asian countries began to receive much attention from the Bank in their efforts to restructure large public sector enterprises and institutionalize liberal systems of governance. Their needs resonated with the Bank's own evolving emphasis of these factors. Hence, substantial resources, about 19.8 percent of the Bank's lending in 1996 alone, were directed to these countries, whereas, in comparison, the entire African continent only received 12.8 percent of the Bank's lending during that same year despite the fact that African nations were the poorest of the Bank's members (World Bank 1996: 248). Finally, the prevailing triumphalist attitude toward the collapse of the Soviet Union – best exemplified by Fukuyama's phrase "the end of history" – deepened the Bank's conviction that its policies during the 1980s and early 1990s were indeed appropriate and effective in promoting development in the Third World.

The Treasury Department's 1982 report on US involvement with aid agencies (discussed earlier in the chapter) recommended that multilateral development agencies should pay greater attention to "market signals and private sector development." Partly in response to American pressure, the World Bank began to pay increasing attention to the private sector in the 1980s and 1990s. In 1987, the Bank created a twelve-member task force called the Private Sector Development Review Group, which consisted of Bank staff and business representatives. The task force was chaired by J. Burke Knapp, a former senior official at the Bank. In its July 1988 report, the Task Force called for improvements in the business environment, privatization, and reform in the financial sector, focusing mainly on creating an environment in which a strong private sector could function. As a result, subsequent Bank lending programs supported regulatory frameworks and changes in policy that encouraged private sector activity, financial sector development, micro-enterprises, and export sector development (World Bank 1989c).

Consistent with this trend, the Bank's urban programs began to acquire a wider focus and emphasized fiscal reform and private sector housing finance. This discussion has tried to show that the broader developments outlined in this chapter were more instrumental in bringing about the wider sectoral and macroeconomic focus than simply technical evaluations of the efficacy of project-based approaches alone.

Implications for the World Bank's urban policy

The trends discussed in this chapter greatly influenced the direction taken by the Bank's urban lending program. As it was always considered somewhat of a "maverick unit," there were many attempts during the Clausen years "to bring the Urban Division closer to standard Bank ideologies."[41] In response, Anthony Churchill, Jaycox's successor as head of the Urban Division, sought to reduce the scale of poverty-directed lending and focus instead on the entire urban economy:

Throughout the development of the Bank's lending program, there has been no confusion on our part that poverty alleviation is only one among many policy objectives.[42]

A city is like a giant oil tanker: you have to turn the wheel a long time before it will change direction. It is all very easy to go in and do some small things, but it is very minor until you begin to move the city as a whole.[43]

As a consequence, the urban programs of the World Bank shifted gears from the low-income housing and residential infrastructure approaches of the 1970s and placed greater emphasis "on policy reform, privatization, urban management and institutional development and training."[44] While the stated goal of the urban agenda in the post-McNamara period was to "integrate the social concerns of the 1970s with the fiscal discipline of the 1980s,"[45] the social concerns soon evaporated. Only fiscal discipline remained after structural adjustment programs took center-stage, as reflected in the 1991 urban sector report. This meant that the direct provision of shelter and infrastructure services to the urban poor was no longer central to the Bank's urban operations. The general contrast between the Bank's urban agenda for the 1970s and the 1980s is presented in Table 4.3. In sum, the Bank's post-McNamara agenda for urban development involved a shift in

Table 4.3 Changing urban agendas of the World Bank

1970s

Provision of land tenure
Selected trunk infrastructure to connect new areas with existing networks
On-site infrastructure (water, sanitation, roads, drainage, and electricity) often based on communal solutions
Core houses ranging from a simple wall with utility hook-ups to completed buildings
Social facilities such as schools, health clinics, community centers
Financing for the plots, core houses, and self-help building materials

1990s

Decentralize government decision-making and strengthen local governments to make them stronger partners in the urban development process
Encourage the private sector to play a greater role in the financing, construction, and operation of urban infrastructure, especially in the areas of housing, urban transport, and solid waste collection
Enhance the productivity of urban enterprises through better provision of basic infrastructure and through regulatory reforms
Improve the functioning of land and housing markets through better macroeconomic policies, reform of zoning and building regulations
Reduce and rationalize subsidies
Adopt a more systematic approach to deal with urban environmental problems and give special emphasis to safe water and basic sanitation

Based on Ljung (1990).

the developing country government's role from that of a "provider" of public urban services, to that of an "enabler" of the private sector. The government's role was gradually redefined into that of a "coordinator" between the market and self-help groups (World Bank 1991b). The new macroeconomic urban orthodoxy at the Bank began to stress tighter fiscal and monetary policy and private sector involvement in low-income housing provision and promoted the Bank's role as "manager." Thus, "governance" and "management" were also added to the Bank's official lexicon of urban development.

The overarching policy shifts discussed in this chapter are particularly well illustrated by changes in the World Bank's housing agenda, summarized below (World Bank 1993b: 7). First, the Bank encouraged governments to play an "enabling" role – that is, they were to move away from producing, financing, and maintaining housing, and work toward improving housing market efficiency. Second, Bank housing assistance acquired a sectoral, rather than a project-by-project, focus. Third, the Bank agreed to assist institutions that have regulatory roles in housing, but left the actual provision and finance of housing to the private sector. Fourth, the Bank searched for "innovative" models for housing finance, which really meant a shift to private lending as opposed to public sector lending. Finally, the Bank sought greater government commitment to improving the means by which housing data are collected and analyzed in order to assess housing sector performance and improve policy formulation and implementation at the sectoral level. These shifts impacted domestic urban agendas in developing countries, as the next chapter will show, using the case of Zimbabwe.

Chapter 5

Beyond global and local

A critical analysis of the World Bank and urban development in Zimbabwe

The preceding chapters discussed shifts in the Bank's urban lending programs, as well as the various forces that influenced its urban agenda. As shown in the last chapter, by the mid-1980s, the Bank had moved away from a project-by-project approach toward a perspective that examined cities in their national macroeconomic contexts. The Bank argued that the government's role ought to be transformed from that of a "provider" of urban services to that of a "supporter" or "enabler" that served as a liaison between the private sector and self-help groups. World Bank urban financial specialist Bill Dillinger argued that "policy change and non-project lending" ought to constitute the Bank's approach to urban problems (Tuck-Primdahl 1991: 7). Michael Cohen, former chief of the Urban Department, observed that the shift "mark[ed] a departure from the past." According to Cohen, the primary objective of this new policy was to "move beyond isolated projects that emphasized housing and residential infrastructure toward integrated city-wide efforts that promote[d] urban productivity and reduce[d] constraints on efficiency" (cited in Tuck-Primdahl 1991: 1). The strategies of the 1980s called for "greater emphasis on policy, urban management, institutional development and training."[1] With particular reference to housing, the Bank's new urban agenda emphasized market reforms and attempted to restructure local government and official housing agencies' finances according to market principles.

It was during this period of transition and flux that the country of Zimbabwe approached the World Bank about the possibility of an urban development loan. According to Fred King of the Southern Africa Department of the World Bank, Zimbabwe asked the Bank's to consider funding a rapid railway system between the capital city of Harare and the satellite township of Chitungwiza.[2] Preben Jensen, also from the Southern Africa Department, pointed out that that Zimbabwe's request arrived at the Bank just as the Bank's overall urban lending policy itself was shifting from a project-by-project approach to the macroeconomic management of urban issues (discussed in Chapter 4). The World Bank saw in Zimbabwe's invitation an opportunity to test whether it was indeed possible to consider the urban sector as a whole rather than fund specific projects in isolation, as the Bank had done traditionally.[3] To this end, according to Jeff Racki, the Bank sent an urban sector mission to Zimbabwe in August 1981 in order to "get a sense of the priority of issues that the government should be grappling with in the urban sector, rather than just helping it build a railway line."[4]

Using the Zimbabwean experience, this chapter examines the insertion of the World Bank's new urban vision into the national urban development agenda of a recipient country. The first section offers an overview of the new black government's development strategy and urban priorities for Zimbabwe, with a focus on urban housing. After contextualizing Zimbabwe's request for World Bank involvement in the mid-1980s, the chapter identifies the core elements of the Bank's urban programs in Zimbabwe and proceeds to critique the Bank's intervention in that country's urban development. While the World Bank's assessments of its own efforts in Zimbabwe adopt a self-congratulatory tone, claiming that the "highly satisfactory" projects were "an excellent example of privatized housing finance" (World Bank 1995: 12), I shall argue that the Bank did not, in fact, influence the broader policy climate in a manner that improved the lives of the urban poor.

Zimbabwe's post-independence development strategy

Zimbabwe gained independence on April 18, 1980, ending some ninety years of colonial and white settler rule. Under the leadership of Robert Mugabe, the Zimbabwean African National Union (ZANU) won the general election. During the political struggles of the 1960s and the guerrilla war of the 1970s, both major nationalist parties, ZANU and the Zimbabwean African People's Union (ZAPU), committed themselves to redressing the socioeconomic and spatial inequalities caused by colonialism. One specific objective listed in a 1972 ZANU manifesto was that "state power will be used to organize the economy for the greatest benefit of all citizens and to prevent the emergence of a privileged class of any kind." This document went on to declare that "an important factor in class formation is the ownership of property," and that in a "free and democratic Zimbabwe, property as a commercial and exploitative factor will be abolished." Mugabe reiterated this point on many occasions, asserting that after independence "none of the White exploiters will be able to keep an acre of their land" (cited in Nyangoni and Nyandoro 1979: 258). Furthermore, ZANU was ideologically committed to a Maoist form of socialism, which gave priority to rural development.

However, inherited social, economic, and political inequalities, as well as the constraints of the Lancaster House Agreement,[5] seriously limited Mugabe's capacity to achieve any radical transformation of Zimbabwean society, in spite of ZANU's commitment to socialism. The Agreement included two provisions that protected whites' interests: (1) whites would retain twenty seats out of 100 in parliament for ten years; and (2) whites would receive a ten-year guarantee on the inviolability of their private property, preventing the new regimes from redistributing land except on a willing seller–willing buyer basis (Mandaza 1986). ZANU's 1980 election manifesto acknowledged this reality, stating that "in working toward socialist transformation, a ZANU government will recognize historical, social, and other practical realities of the capitalist system which cannot be transformed overnight" (cited in Nyangoni and Nyandoro 1979: 258).

In contrast to the flamboyant rhetoric of the pre-independence period, Mugabe's inaugural address was more subdued and racially conciliatory:

> We will ensure that there will be a place for everyone in this country. We want to ensure a sense of security for both winners and losers. There will be no sweeping nationalization; the pensions and jobs of the civil servants will be guaranteed and farmers would keep their farms; Zimbabwe will be non-aligned. Let us forgive and forget. Let us join hands in amity.[6]

Mugabe's gesture toward the white population indicated his recognition that, even as his government aspired to socialism and an identification with the peasantry, the vitality of productive economic sectors would have to be ensured if other development objectives were to be met. Zimbabwe inherited not only a well-diversified and strong commercialized agricultural sector, but also a well-developed manufacturing sector, both dominated by whites. The new government did not wish to destroy these sectors by provoking an exodus of the skilled white population, which had proved extremely debilitating in neighboring Mozambique (Gordon 1984; Reed 1987).

In 1981, the new government published its first major policy document, *Growth with Equity*, in which it attempted to reconcile the twin objectives of economic growth and social justice. During the same year the government coordinated The Zimbabwe Conference on Reconstruction and Development (ZIMCORD) in order to discuss development strategies and seek international assistance. In 1982, the more comprehensive Transitional National Development Plan (TNDP) was published (Government of Zimbabwe 1982a). While the policy documents and the conference called for an eventual socialist transition, they nevertheless emphasized the role of the private sector and focused on strategies that built on existing economic strengths.

The housing question and government responses

The urban system in Zimbabwe reflects the uneven incorporation of the region into the capitalist world system through colonialism. O'Connor's (1986) typology of African cities characterizes Zimbabwean towns as "European" because their planning and design was determined by, and for, settler needs. Cities like Harare (Salisbury) in Zimbabwe initially emerged as political bases for colonial administration and were established along the main infrastructure routes or along mining corridors (Wekwete 1988) (see Figures 5.1 and 5.2).

In spite of white ruling class efforts to separate the races spatially, there has always been a black presence in white urban Zimbabwe. Generally, the blacks in urban Zimbabwe were tolerated as long as they had jobs in the city as domestic servants etc., but their families were not permitted to move from the tribal trust lands.

Following its Unilateral Declaration of Independence (UDI) from Britain in 1965, white Rhodesia strengthened relations with South Africa and attempted to install rigid apartheid-like segregation.[7] However, in spite of the UDI and legislation intended to keep blacks out of urban areas, African urbanization continued, increasing to 15 percent in the 1970s (Patel and Adams 1981). The figure increased in the later years of the Liberation War (mid-1960s through 1980), and following independence there has been a steady increase in rural to urban migration.

Early housing for black single male workers in Zimbabwe's urban areas

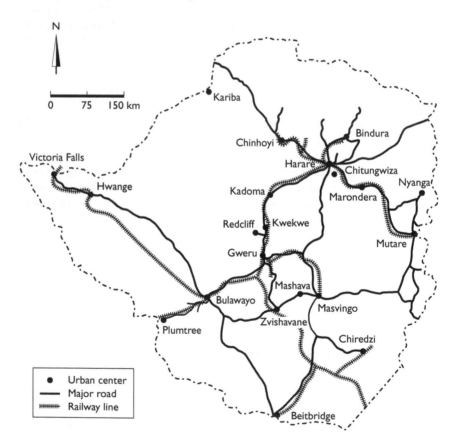

Figure 5.1 Major urban centers of Zimbabwe.

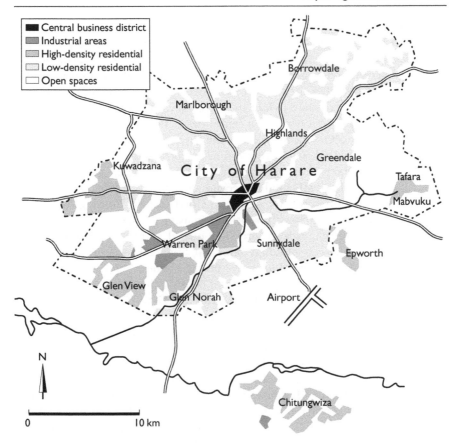

- ■ Central business district
- ▨ Industrial areas
- ▨ High-density residential
- ▢ Low-density residential
- ▢ Open spaces

Borrowdale

Marlborough

Highlands

Greendale

Kuwadzana

City of Harare

Tafara

Mabvuku

Warren Park

Sunnydale

Epworth

Glen View

Glen Norah

Airport

N

Chitungwiza

0   10 km

Figure 5.2 Map of Harare.

was located either on employers' premises or in barrack-style hostels at the periphery of towns. It was not until 1941 that the Land Apportionment Act was amended to provide for the establishment of urban townships for blacks. Although government policy aimed to discourage blacks from permanently residing in urban areas, the increasing influx of blacks and the growing demand for their labor created a strong countervailing force. Over time, an increasing number of blacks were becoming permanently urbanized, and their children had little or no direct contact with the rural areas (Möller 1974).

As the white settler regime came to terms with the permanence of African urbanization, small dwelling structures – usually one- or two-bedroom, matchbox-style houses – were built for the African population. In the early 1960s, the white governments experimented with the idea of core housing projects for formal sector wage earners. This initially posed a dilemma for the white government because core projects were based on the idea of home ownership and Africans were not allowed to own land in urban areas. Eventually, home ownership was legalized for some Africans, and in the township of Glen View, for example, more than 7,000 plots with wet cores were constructed (Zinyama *et al.* 1993). However, the provision of "married" accommodation continued to lag behind "single" houses until the early 1970s (Patel 1984). In an attempt to deal with the burgeoning African urban population in the mid to late 1970s, the government erected a number of ultra-low-cost housing units on a rental basis. The size of these dwellings ranged from 30 to 60 m². Sometimes the kitchen was only a roofed area outside; toilets and bathrooms were either inside or outside.

Zimbabwe in general, and Harare (the capital city) in particular, inherited a housing system that embodied the economic, political, social, and cultural asymmetry that is inherent in a settler colonial society. Housing for whites was generated in a capitalist, differentiated housing market based on freehold tenure. Most white homes were privately owned and funded through loan capital generated by conventional capitalist financial institutions, such as building societies and banks. Housing for blacks, on the other hand, was mainly based on public land ownership and rental housing that was administered as a separate entity by local authorities (Davies and Dewar 1989).

With respect to the overall housing stock for blacks in Harare at the time of independence, there were about 70,000 houses under city council control and 29,000 in Chitungwiza. There were also about 35,000 units for domestic workers, 5,000 units in hostels, close to 3,000 units in Epworth, and about as many in transitional camps in Chitungwiza (Hoek-Smit 1982). This housing stock was heavily crowded because of the large number of lodgers in addition to the officially registered tenants. During the last years of the Liberation War, the government was unable to stop the influx of the African population into the urban areas, and a number of squatter settlements began to grow in spite of the government's firm anti-squatter policy. Thus, at independence, the new black government inherited a rapidly growing, rigidly segregated capital city. The existing stock of low-income housing was severely overcrowded, and squatter areas were expanding.

The ruling ZANU party, given its Maoist leanings, did not feature urban development as prominently as rural development in its transitional development plans. However, the government did make a commitment to address the lack of housing and sought to meet the basic needs of the urban population. For example, during the ZIMCORD conference, the government committed itself to overcoming "the acute housing shortage" and planned to "implement a program to build 167,000 low-cost houses in rural and urban areas over the next five years" (Government of Zimbabwe 1981: 58). This was reiterated in the TNDP. Eddison Zvobgo, Minister of Local Government and Housing, affirmed that:

> The first democratically elected Government of the Republic of Zimbabwe has vowed to assure all its people of shelter, food and clothing. These

need not be elegant; it is enough that they are of the kind that permit human dignity and human life.

(Zvobgo 1981: ix)

During the transitional period of the early 1980s, housing policy was encumbered by several major organizational and financial constraints. First, housing was shifted from one ministerial portfolio to another; at first, it was designated the responsibility of the Ministry of Local Government and Housing, but in 1982 this ministry was subdivided in order to establish a separate Ministry for Housing. However, just as it began the work of investigating and addressing housing concerns, this separate ministry was combined with the Ministry of Construction to form the Ministry of Construction and National Housing.

The Ministry of Construction and National Housing (MCNH) administered two funds, the National Housing Fund (NHF) and the Housing Guarantee Fund (HGF), for public sector-supported housing. The NHF was the main source of finance for urban low-cost housing, providing thirty-year loans that were disbursed at an interest rate of 9.75 percent. Initially the NSF relied solely on the central government for its resources. However, by the mid-1980s it was supplemented by World Bank and USAID funding. As the NHF had always been plagued by a shortage of funds, development assistance constituted about 60 percent of its funding portfolio in the early 1990s. The HGF provided house purchase guarantee schemes, by which the central government guaranteed repayment for a privately financed mortgage. This scheme was mainly for civil servants (Mutizwa-Mangiza 1991; Rakodi 1995).

Second, while various political parties agreed in 1980 that Zimbabwe faced a housing crisis, their opinions regarding the extent of the crisis varied. The new government frequently opted to use conservative numbers to estimate housing demand and shortages, thereby downplaying the problem. For example, the 1982 census reported Harare and the satellite city of Chitungwiza as having populations of 656,011 and 172,556 respectively. However, University of Zimbabwe sociologist Diana Patel (1984) disputed these figures, estimating Harare's population to be closer to 1 million and that of Chitungwiza to be 350,000. In fact, in a USAID report, Hoek-Smit (1982) estimated 650,000 as the population of Harare's high-density black townships alone. As Patel points out, "in spite of the disagreement over the official numbers, the government's estimate was used as a basis for formulating policy and defining the level of housing need. These grossly underestimated the problem."[8] The government's estimate for Harare, based on census figures and the official waiting list, is a case in point: while Hoek-Smit (1982) puts the demand around 60,000, the official estimate was 20,466.

Despite such debates over numbers, the government did agree that the housing shortage was a critical problem that demanded attention. It mapped out a housing policy which featured four primary issues: (i) the promotion of home ownership, (ii) the establishment of new minimum housing standards, (iii) aided self-help, and (iv) the enforcement of a strong anti-squatter stance. Such a policy indicates that the new government chose to revise the housing policies of the colonial regime rather than pursue a radical departure from them.

During the liberation struggle, Zimbabwe's major liberation movements called for the abolition of private property and the redistribution of white-owned land. However, the pragmatic post-independence posture of the new government prevented these goals from being implemented. Instead, ZANU actively promoted home ownership for the black population, as seen in Zvobgo's statements:

> It is intended that all new housing developed in our Local Government Areas [formerly African Townships] will be available for home ownership.[9]

Zvobgo expressed similar sentiments at an Urban Councils conference:

> Ownership, and more particularly pride of ownership, is the key to the improvement of the way of life of the majority of our people.[10]

As a result of the central government's commitment to home ownership, tenants were allowed to buy previously rented dwellings. For each of the first five years of tenancy, tenants were granted a 2 percent discount, which increased to 3 percent for each year of a further ten-year period, and to 4 percent for each year of a subsequent fifteen-year period. Hence, a tenant dwelling on a property for thirty years would receive a discount equivalent to 100 percent (Teedon 1990).

The new black government sought to upgrade the minimum housing standards set by the old regime. As noted in the previous section, the colonial government promoted the ultra-low-cost core housing unit as a cheap option for African urban residents during the waning years of its rule. While the construction of ultra-low-cost units continued in Chitungwiza between 1980 and 1982, these units were increasingly criticized by post-1980 politicians, who saw them as an enduring symbol of the colonial period. Thus, in October 1982, a new minimum standard for plots and houses was formulated by the Minister of Housing (Government of Zimbabwe 1982a):

- No plot should be smaller than 300 m².
- Whenever possible, larger plots of 300, 400, 500, and 600 m² should be incorporated into future projects.
- The minimum core house should consist of two bedrooms, a dining-room, a kitchen and a toilet/shower.

- The minimum floor space should be 50–60 m^2.
- Building permits should not be granted for dwellings built to lower standards.

Initially, this call for higher standards in housing was met with popular support, as it was perceived to be a reversal of the old colonial order. However, when applied with the principle of full cost recovery, the new standards created adverse consequences for low-income groups. As a result, appropriate housing standards became a major issue of contention by the time the World Bank got involved in urban housing projects in Zimbabwe. In an attempt to make these new higher standards affordable for the population, the government encouraged and promoted self-help building and the idea of "building brigades."

The self-help strategy promoted by the new government was an aspect of its wider housing policy supporting home ownership. The idea was first promoted in the early 1970s by the former settler regime, especially in the Harare township of Glen View.[11] The government saw the Glen View scheme as a success, and worked toward a similar scheme in Kuwadzana, a planned, low-income township in Harare. This project, partly funded by the USAID, is discussed in detail later in this chapter.

To complement these self-help schemes, the new government actively encouraged the organization of building brigades. Government officials who had toured the developing world in early 1981 were shocked by the proliferation of large-scale squatter areas in many countries and hoped to prevent such settlements in Zimbabwe. They were, however, impressed with Cuba's solution to its housing problems, especially in the suburbs of Havana. Cuban officials proudly claimed they had conquered their housing shortage through building brigades. Following the Cuban example, the Ministry of Housing subsequently adapted the idea of building brigades for Zimbabwean conditions. Zvobgo extolled the virtues of building brigades:

> We propose to establish "people's construction brigades" across the country. Any family in need of but unable to afford a house will have to join a brigade. Together the brigade members will donate their labour. The State should make available the materials and avoid leaving the public to the whims and appetites of private enterprises in this regard.
>
> (Zvobgo 1981: x)

The Ministry sketched out ambitious plans for building brigades to produce over half of the housing that had been planned for construction. However, the plans were far from successful, and very few self-help homes were actually built through this mechanism; in the Kuwadzana project, for example, only fifty or so homes were built by the brigades. Eventually, small private contractors came to be used to implement the self-help schemes.

In light of the failure of building brigades and its attempt to maintain standards and implement its pro-rural development strategy, the new government adopted a harsh anti-squatter policy. For instance, almost immediately after independence, it resorted to bulldozing several squatter settlements around Harare. The Ministry of Local Government and Town Planning at the time summed up its stance as follows: "anyone who intends to squat anywhere must forget this because we are going to be ruthless."[12] For example, the government showed that it "meant business" in the Harare township of Mbare Musika in 1980, where a squatter camp was demolished and inhabitants were exhorted to return to their ancestral rural areas. However, very few squatters actually did so because the rural areas lacked economically viable employment. While some families were accommodated in the extremely overcrowded conditions of former "single" men's hostels (Patel 1984), the vast majority of them remained homeless. Chirambahuyo, another large squatter settlement in Chitungwiza, was also demolished by the new regime[13] as part of the government's endeavor to enforce its minimum standards requirements and its policy of rural socialization. The government was often callous and naive in its expectations regarding the resettlement of residents. For instance, the Minister of Housing at the time, obstinate about the supposed advantages of resettling squatters, is said to have expressed disappointment when survey results indicated that only 700 out of 3,000 Chirambahuyo residents wished to return to rural areas (Patel and Adams 1981).

The government's bulldozing policies in Chirambahuyo destroyed the poor's valuable alternative housing stock. Z$112,670 is a conservative estimate of the value of the demolished self-built houses, a substantial investment made by the residents (Patel and Adams 1981), for which they were not compensated. Destruction of this settlement also meant the destruction of an extremely viable and vibrant community. Through various self-help efforts, the squatters of Chirambahuyo formed resident associations, which, in turn, linked with the Chitungwiza Urban Council, where they were given representation. Through such linkages, residents fought for improved community facilities. In a detailed study of the settlement, Patel and Adams (1981: 87) concluded that "clearance of the area would destroy what people have built up both physically and socially." Nevertheless, the Chirambahuyo settlement was bulldozed in May 1982,[14] and only some residents were offered low-cost housing at monthly payments of Z$14. Very few, however, could afford this sum, and, consequently, many residents resorted to squatting elsewhere, particularly in the Mayambara settlement in Chitungwiza. Despite the lack of adequate, affordable housing for the urban poor, the government persisted in its squatter demolition campaign. In October 1983, with "Operation Clean-up," the government launched a nationwide program to eradicate squatter settlements (Schlyter 1990).

That same year, the government grudgingly decided to upgrade the squatter settlement of Epworth,[15] where some 28,000 people lived in substandard housing. After a USAID-assisted feasibility study was conducted, the government provided some Z$2 million for the upgrading scheme, which included water reticulation, sewerage systems, road networks, and bus routes (Butcher 1986). However, the government's reluctant decision to upgrade the settlement included a prohibition against expansion beyond the boundaries that had been identified in 1983 using aerial photography. Thus, when the upgrading program actually commenced in 1985, all informal units built after 1983 were destroyed. From this, it is clear that the Epworth upgrading scheme did not reverse the government's general anti-squatting policy.

International agencies and urban development

The previous discussion showed that at independence tens of thousands of homeless people resorted to lodging and squatting, while many more could not afford even the most basic accommodation. A key objective of the post-liberation TNDP had been to "ensure adequate housing and related services at affordable prices for all, irrespective of geographical location or socio-economic group." Despite such pronouncements, however, the victorious nationalist war of liberation meant very little in terms of immediate, tangible gains for the urban poor in Zimbabwe. Patel (1984: 194) characterized the government's initial housing programs as barely existent, owing to a "four year hiatus ... largely brought about by the lack of a clearly thought out housing policy." The TNDP (1982) stated that the urban backlog could be eliminated in eight years if 115,000 units could be constructed at a total investment cost of Z$525.2 million. However, in reality, housing construction and investment fell far short of the plan. During the first two years of the TNDP, only 7,500 houses were built and only 12,500 serviced stands were constructed. By the end of 1985, only 13,500 houses out of the 115,000 planned total had been completed. Budget allocations to the Ministry of Construction and National Housing fell far short of what was required to complete construction. In 1985, for example, although it was estimated that the Ministry needed Z$325 million to construct the 57,500 units planned, only Z$123.7 million, or 38 percent, of this amount was budgeted (Potts and Mutambirwa 1991).

As a result, the goals of the TNDP were not met and the housing backlog continued to increase during the mid-1980s. In 1985, for example, the backlog was reported as 110,000 units, with 63,000 of those in Harare and Chitungwisa alone (Mutizwa-Mangiza 1986). While the government's vicious campaign of squatter eradication continued to feed the illusion that Zimbabwe did not suffer from a low-income housing problem, Mafico observed:

> If you go into the low-income areas, you won't see squatter settlements. Visitors to Zimbabwe say that "this is very nice for a developing country." But the problem is hidden. I did a survey for my research of house

occupancy and found that in houses designed for three to six people you frequently get 25 to 30 people living in them. This is what is actually happening. You have a problem that is hidden from the public eye.[16]

While official policy continued to insist on high standards of building design, the vast majority of the low-income population could not afford even the cheapest of these models.

By the mid-1980s, the Zimbabwean government admitted to a shortage of funds to address growing demands for low-income urban housing (Government of Zimbabwe 1985). In 1985, reconsideration of the crisis in light of existing housing policy led the government to seek the assistance of international development agencies. In 1986, the government presented the First Five Year National Development Plan, which stressed the importance of fostering public–private partnerships. The plan aimed for (Government of Zimbabwe 1988):

- the mobilization of funds by the public and private sectors for housing construction;
- the encouragement of building societies to play a greater role in the provision of low-cost housing;.
- the facilitation of self-help approaches by increasing the use of small contractor/construction companies in the building of houses;
- the introduction of tax exemptions for investments in rural growth points to reduce rural to urban drift.

That same year, the tenth Conference on Housing and Urban Development in Sub-Saharan Africa (CHUDSA) was held in Harare. This conference, like the First Five Year National Development Plan, also promoted public–private partnerships in housing and urban development. Robert Mugabe, then prime minister of Zimbabwe (now president), addressed the conference with the following opening remarks:

> Housing continues to be one of the most neglected aspects of our development efforts in the Third World and in other countries the world over . . . We must therefore take action to give priority to housing as an integral aspect of our national development initiatives.
>
> Public and private sector partnership in housing and urban development is both significant and timely . . . [the Zimbabwean government] recognizes the need for this partnership.

> (Mugabe 1986: 3–7)

Mugabe's call echoed the evolving philosophy of the World Bank and USAID at the time, which advised a reduction in direct economic intervention by the government. It was within this climate of compatibility that the World Bank entered and impacted the formulation of low-income housing policy in Zimbabwe.

In the absence of a domestic solution to the low-income housing crisis during the mid-1980s, central and local government in Zimbabwe joined forces with USAID and the World Bank to map out a strategy. The emerging policy emphasized a market-oriented approach and called for private sector involvement in financing low-income housing. With the encouragement and support of USAID, Harare began expanding its aided self-help schemes in the mid-1980s.

As noted earlier in the chapter, the government intended to provide housing for low-income households in Kuwadzana township, Harare, on a full cost-recoverable basis. It was assisted by a US$38 million grant from USAID. Between February 1984 and December 1985, 6,000 plots of 333 m^2 were assigned to families on Harare's waiting list. Families on the waiting list were required by the government and USAID to have a maximum income of Z$175 (Rakodi and Mutizwa-Mangiza 1989). Under this aided self-help scheme, allottees were required to build four rooms and an ablution facility within eighteen months. USAID tried, unsuccessfully, to encourage the government to lower the plot size in order to allow more houses to be built, but the government resolutely refused to lower standards. (Standards, discussed later, became a major point of contention within the Bank's first urban project.)

Previous government housing schemes had promoted home ownership based on cost recovery principles, but the Kuwadzana project involved the private sector for the first time, on a pilot basis. USAID courted Beverly Building Society to finance 10 percent of the homes for this project, and the central government provided a guarantee for Beverly's loans on recommendation of USAID (USAID 1985). Through the project, Beverly allotted thirty-year loans of Z$2,500 (27.5 percent of the allottee's income) designated for the construction of a four-room core house within eighteen months (Mafico 1985; Schlyter 1985).

Initially, as the town hosted some 2,000 independent builders, Kuwadzana experienced an impressive increase in entrepreneurial activity and house delivery. According to Harare's Housing Director, the sites were "a hive of activity and energy. The initiative is to be commended" (Harare 1985: 12). However, Harare's lowest-income quartile, unable to afford the houses, did not experience the benefits of the project (Schlyter 1990); 80 percent of the lots went to individuals whose incomes were near the upper limit of Z$175, although the mean income for Harare was only Z$130, and nearly one-third of the population earned less than Z$70. Therefore, Schlyter (ibid.: 215) concluded that it "would be more honest to call Kuwadzana a middle-income housing project in which the low-income population come in as lodgers." In addition, the estimated cost to complete a four-room core house in 1985 was Z\$9,000; with government loans of only Z\$2,500 per family, low-income families were unable to complete the project (Butcher 1986). Also, the Kuwadzana housing schemes became particularly vulnerable to absentee landlordism and raids by higher-income groups (Rakodi 1989), which became more widespread over the years, especially during the World Bank's later programs.

By encouraging home ownership and the involvement of local building societies in project finance (albeit with a government guarantee), the USAID project set the stage for private sector involvement in addressing Zimbabwe's low-income housing needs. Thus, the participation of building societies in low-income housing finance also became the cornerstone of the World Bank's subsequent urban programs.

The World Bank's urban programs in Zimbabwe

The World Bank's urban sector mission to Zimbabwe produced a detailed report on the urban condition in the country, which recommended a series of policy changes and repeatedly commented on the need to address "the extreme duality of Zimbabwe's economy" (World Bank 1985b).[17] For the Bank, the solution to the duality was to provide low-income areas with the same kind of institutions and services that were available in the middle- and high-income, areas. In essence, this strategy entailed increasing home ownership and utilizing the private sector to expand market-oriented housing delivery. The Bank's core programs in Zimbabwe thus revolved around creating "an enabling environment" for the private sector to finance housing projects (World Bank 1993a: 1). The Bank deemed that this would be the most effective way to address urban problems in post-independence Zimbabwe.

With particular reference to urban housing, the mission recommended a focus on supply, given limited public resources and rapidly growing demand. It stressed that the old regime's low-income housing policies ought to be carefully examined for their merits, and not discarded simply because of their association with the colonial past. According to Jeff Racki, task manager of the Bank's first urban project in Zimbabwe, the mission was concerned that the new government's resentment and distrust of the former regime's policies would cause them to "lose the phenomenal capacity of the local government system to address housing and infrastructure needs."[18] The mission was aware of three major limitations of the old regime's housing programs (World Bank 1985b: 55): (i) they were based on segregated settlement patterns and disadvantageous, inefficient locational policies; (ii) the housing strategies did not foster beneficiaries' participation in program formulation; and (iii) the system of racial discrimination contributed to an inefficient state bureaucracy with separate administrative units for the different racial groups.

Despite these problems, the Bank believed that the past programs contained major strengths, such as (i) an emphasis on local authorities' autonomy and capacity to identify, prepare, finance, and implement housing projects; (ii) a strict policy of replicability, requiring full costs from the beneficiaries; (iii) careful attention to the design of affordable alternatives; and (iv) a shift from the provision of rental housing to home ownership. The urban mission advised that these "positive features" from the old regime ought to be complemented with strategies identifying new private sector sources for funding low-income housing; this, it felt, would help supplement the already strained budgets of local authorities.

The mission also identified four low-income housing options for Zimbabwe's cities: standard housing units, core units, ultra-low-cost units, and sites-and-services units. According to the Bank's calculations, three-quarters of urban households could not afford the standard unit, but half could afford the core housing unit and four-fifths could afford the ultra-low-cost unit (World Bank 1985b: 63).

Based on its own assessment of past housing programs and available options, the World Bank's urban sector mission to Zimbabwe concluded that (ibid.: 32):

- The government needs to recognize impending urban influx, which is likely to be much larger than the towns or the central government expect.
- The financial health of the cities should be preserved through sound fiscal management by the local governments, with support from the central government.
- Attention must be paid to the manpower needs of local authorities. The benefits of maintaining well-run, autonomous, self-sufficient cities capable of delivering and operating the full range of local services must be emphasized.
- The pitfalls of delivering high standard services to a few and leaving many with no services should be avoided.
- Given the post-independence reality of rapid urban population growth and limited public sector funds, past practices in urban service delivery could not be sustained.
- Urban problems could be controlled if government activity in the housing and infrastructure sector shifted to providing serviced land and involving the private sector (building societies and households' own savings) in financing housing construction.

When discussing this report with Zimbabwean authorities, the Bank also stressed the need to strengthen local government authority and capacity, and called for "prudent policy weighted to cheaper options" (ibid.).

Urban I

Effective on June 7, 1985, the Bank's first urban development project in Zimbabwe, called Urban I, set out to implement the recommendations of the urban sector report discussed above. With respect to low-income housing, the main objective of the project was to "increase the supply of affordable housing and related services to large segments of the poorer population, and to improve the system of housing finance" in four major cities: Harare, Bulawayo, Mutare, and Masvingo (World Bank 1991c: 8) (see Figure 5.1). Urban I aimed to (World Bank 1984b: 9–10):

- Promote the financial and institutional capacity for increasing the supply of affordable housing and related services in Zimbabwe by restructuring the housing delivery and mortgage financing markets.
- Strengthen the government's efforts to preserve, through manpower development and training programs, both the fiscal integrity as well as the technical capacity of the local authorities for supplying, operating, and maintaining essential urban infrastructure and services.
- Provide strategic support in key areas of urban development where new policy directions (in addition to those areas described above) are currently being formulated, such as transportation.

The total project costs were estimated to be about US$112.5 million, of which US$43 million, or roughly 40 percent, was to be funded by the World Bank. More than 90 percent of the project's funding went to the housing sector. The financial breakdown of the project costs and the amounts contributed by the co-funding agencies are depicted in Tables 5.1 and 5.2 respectively.

World Bank and Zimbabwean government funds were to be made available to the NHF, which was part of the MCNH, and the Central Development Loan Fund (CDLF), which was part of the Ministry of Local Government and Town Planning (MLGTP). The flow of funds for this project is depicted in Figure 5.3. Funds for general infrastructure were to be lent to the local authorities through the NHF, whereas funds for community development facilities, technical assistance, transportation improvements, and the establishment

Table 5.1 Components of Urban I

Components	Total cost (US$m)
Site development and servicing	57.4
Transport	3.4
Institutional development	7.2
Housing finance	49.2
Front end fee	0.1
Total	117.3

Source: World Bank (1984b: 17).

Table 5.2 Financing structure of Urban I

Organization	Contribution (US$m)
World Bank	43.0
Zimbabwean government	22.4
Building societies	37.6
Commonwealth Development Corporation (CDC)	9.6
Total	112.6

Source: World Bank (1984b: 17).

of the Zimbabwean Association of Accounting Technicians (ZAAT) were to be lent through the local authorities and the General Development Fund (GDF).

The project specifically tried to address three problem areas, which, from the Bank's perspective, made the provision of low-income urban housing difficult in Zimbabwe: (i) excessively high standards for low-income housing, which the Bank believed were exacerbating the housing crisis; (ii) the absence of private sector involvement in generating resources for low-income housing supply; and (iii) the absence of linkages between the public sector housing program and private sources of finance.

The project intended to provide some 12,000 serviced residential plots, of which 70 percent were to be 300 m^2 and the remainder larger. Off-site infrastructure costs (water, electricity, sewerage, and waste management) were to be recovered through tariffs and rates/supplementary charges (for main roads and street lighting), while on-site infrastructure costs were to be recovered from plot sales (land, local roads, surface drainage, and survey) and tariffs (water and electric reticulation). The minimum building requirement was the construction of one room with wet core, to be completed before the plot was occupied, and four rooms to be built within eighteen months.

One of the major disagreements between the Bank and Zimbabwean government officials involved the issue of appropriate design standards for low-income housing (World Bank 1994b). During pre-independence years, the white government promoted the "ultra-low-cost house" as a residential dwelling that would be within the reach of the lowest-paid workers. The ultra-low-cost houses were designed to be the cheapest structures available and consisted of one room with ablution facilities. The new government regarded this form of shelter as "inconsistent with post-independence expectations" (Underwood 1986: 30). As discussed in the previous section, upon coming into power in 1980, Mugabe's government officially established four-roomed dwellings as the minimum standard for those earning below Z$650/month (Musanda-Nyamayaro 1993). Through such standards, the new government attempted to demonstrate its rejection of the old regime's standards, especially after having criticized them throughout the liberation struggle. As a result, the government regarded the Bank's "new and improved" 1984

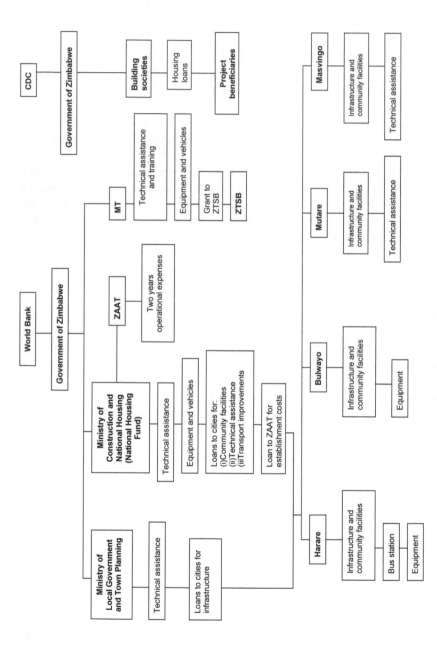

Figure 5.3 Flow of funds for Urban I. Source: World Bank (1984b: 19).

housing standards to be unacceptable replications of the old colonial order. The issue of appropriate standards was a topic of fierce debate between Zimbabwean officials and both the World Bank and USAID. In reference to the government's insistence on the four-room core unit, the USAID Country Development Strategy Report notes that:

> Such a policy . . . [has] . . . profound costs and subsidy implications . . . which would make it impossible for several donors to continue with low-cost housing programs, all of which are based on the concepts of full cost recovery, affordability to low-income target groups, and the use of private and aided self-help construction techniques.
>
> (USAID 1984: 18)

For almost a year, extensive discussions continued. According to Racki, Zimbabwean officials felt that the "Bank was strong-arming them" by insisting on lower standards for low-income housing.[19] Eventually, a compromise was reached between the government and the Bank: while the four-roomed house would be regarded as an ultimate goal, people could build smaller houses and enlarge their dwellings as resources became available. Thus, an incremental approach to meeting the government's minimum standards was adopted.

During the period of white settler colonial rule, local government financed housing schemes at a subsidized rate for a small proportion of the urban African population. According to Racki, the new black majority government faced political pressure to continue to offer and expand such schemes as a result of this precedent. From the Bank's viewpoint, these schemes ran the risk of stretching the fiscal resources of the local and central state, and "were not sustainable in the long run," in spite of their political appeal.[20] Thus, the Bank's urban sector mission cautioned against them, concerned that the government might actually implement such schemes.

On the government's part, subsidized public sector housing programs were never considered seriously except during the liberation struggle. As pointed out in the previous section, the government tried to promote home ownership and cost-recoverable sites-and-services schemes during the mid-1980s, but found that it was still unable to meet growing housing needs. Following a report by the Ministry of Community and Cooperative Development that "housing demand will continue to rise and the public sector has limited resources of finance and manpower which it can provide to housing" (Government of Zimbabwe 1987: 6), the government began to seek international and private sector assistance to address the low-income housing problem. Thus, in concert with the government's call for more public–private partnerships, the World Bank also looked to the private sector for alternative funding for low-income housing and expanded the role of private building societies in its projects.

As discussed in Chapter 4, the Bank's urban programs of the 1980s were characterized by a shift away from project-based approaches toward a broader macroeconomic focus. Through such a framework, the Bank sought to create an "enabling environment," as noted earlier, for private sector involvement in housing finance, even for low-income groups (World Bank 1993a,b). The Bank tried to incorporate this thinking into Urban I by involving local building societies. Although these societies were the main source of mortgages for middle- and upper-income applicants, the Bank sought their participation to cover the capital costs of plot and housing construction for low-income groups. Ultimately, the Bank not only persuaded building societies to finance low-income housing on a scale wider than previous efforts, but persuaded them to do this without a central government guarantee. By involving the building societies, the Bank aimed to bring together two separate systems of housing finance delivery that had existed in Zimbabwe prior to independence: it sought to integrate private sector-based financing options that served middle- and upper-income households with the public sector-based financing delivery options intended for low-income households. It is important to note here that this aim was consistent with the Bank's general, post-1970s urban agenda that sought to "treat the housing sector as a whole" (World Bank 1984a: 5).

Initially, the Bank found it difficult to persuade the building societies to fund low-income housing on a non-subsidized basis. As Racki, Urban I's task manager, recalls:

> They weren't exactly jumping up and down to fund low-income housing. The old codgers who ran the building societies were part of an old regime and their mindsets were not going to shift very easily.[21]

Before independence, building societies did not finance urban low-income housing because of the white regime's exclusion of blacks from urban areas. After independence, despite the end of official segregation, building societies still did not enter the low-income housing market immediately. Patel attributes their hesitation to a number of factors: they did not have a history of making loans to low-income home buyers; standards in high-density areas were lower than what was traditionally acceptable to building societies; the cost of administering a large number of small loans was high; and they feared a high default rate. For all of these reasons, building societies considered low-income housing highly risky.[22]

In the end, four factors prodded the building societies to finance low-income housing: (i) political criticisms that they borrowed from the poor to fund the houses of the rich; (ii) the middle- and high-income housing market stagnated after independence; (iii) international development agencies offered building societies financial assistance to computerize; and (iv) the government offered them a financial incentive by permitting them to issue tax-free "permanent paid-up shares" (PUPS) at 9 percent (explained below).

First, with respect to the political criticism, Nyoni observed that the new government and the general black population resented building societies because they were "typically the place where most of the Black labour force put their savings but were never given access to loans because the money went toward low-density White housing."[23] The building societies were accused of racism, conservatism, and a refusal to accommodate the poor (Mafico 1991; Kamete 1999). The Deputy Minister of Local Government and Town Planning accused building societies of various forms of discrimination in the past and threatened to "strike any institution off the register if it showed racial prejudice in granting loans."[24]

In light of the new regime's socialist leanings, such criticisms led to rumblings in government corridors about the merits of nationalizing the building societies. While the government did eventually adopt a pragmatic approach to politics and development after independence, "the building societies were a prime target for nationalization given their racist history."[25] When Robert Mugabe called for public–private financing for low-income housing at the tenth CHUDSA (mentioned earlier), he added that private enterprises were obligated to assist social development in Zimbabwe because of their exploitative practices in the past:

> It is common knowledge that wealth is generated through the exploitation of natural resources and sub-Saharan Africa is endowed with these natural resources . . . after independence, the private sector continued to dominate and control these post-independent economies. It therefore follows that the private sector has an inescapable obligation and a social responsibility to invest in housing and urban development.
>
> (Mugabe 1986: 3–7)

The building societies, in turn, were aware of these criticisms and realized the expediency involved in funding low-income housing. As the housing projects manager for the Central African Building Society (CABS) put it:

> The building societies appreciated that they could not remain on the fence or gain the reputation of being borrowers from the poor and lenders to the rich. For the sake of their own survival they had to get in on the low-income housing act.
>
> (Beresford 1992: 580)

Second, the middle- and high-income housing market declined just after independence; in fact, from the mid-1970s, the higher-income housing market began to slump as the Liberation War escalated and whites left the country. Blacks at this time had not accumulated sufficient wealth to purchase the emigrants' homes (Bond 1993). A few townhouses were the only new residential buildings under construction at independence in the high-income–low-density suburbs. Except for the residences of senior diplomats, which experienced high demand as a host of embassies and non-governmental organizations moved into the nation's capital,[26] there was virtually no demand for high-income properties. As a spokesperson for the Association of Building Societies in Zimbabwe observed:

> At one time, investment in [low density suburbs] ran to almost 40 per cent of our assets. But now this has dropped to less than 10 per cent and most of this is not for new houses but for renovations and extensions on existing ones.[27]

The collapse of the middle- and high-income housing market "helped to convince building societies to finance low-income housing because they weren't lending to anyone."[28] The World Bank sector report on urban development in Zimbabwe echoed this observation:

> A case could also be made for involving the building societies in low-income housing simply to preserve their dynamism during the period of a depressed upper-income housing market. At present, there is no sign that the slump is abating and it may well last for years. It is not clear that the building societies could survive, as mortgage lending institutions, without entering the low-income housing field in one way or another.
>
> (World Bank 1985b: 81)

The Bank presented this situation to the government and the building societies as an opportunity to enter into a mutually beneficial partnership.

> The government gains if the potential dynamism of the societies can be exploited for low-income housing; and the building societies gain if they can enter this fast growing field while their main historical business is severely depressed.
>
> (ibid.)

Third, the building societies were offered a grant to computerize and streamline their operations, since they had claimed that participating in Urban I would phenomenally increase the volume of loans, which would then require computer processing. Therefore, as a condition of their partnership, building societies wanted the Bank's project loan to cover the cost of computerization. The Bank was amenable to this request, but, according to its Articles of Agreement, it could lend money to the private sector only through governments. Thus, the Bank presented the building societies' request to the Zimbabwean government for consideration. The Zimbabwean government, at first, did not understand why it had to borrow from the World Bank to subsidize the computerization of a private enterprise and its reluctance almost jeopardized the proposal. In the end, the CDC awarded the building societies a US$9.6 million grant for computerization.

Finally, the government reluctantly agreed to offer the building societies a financial incentive by permitting them to issue tax-free PUPS at 9 percent, with a ceiling investment of Z$75,000 for individuals and Z$35,000 for companies. The Bank hoped that PUPS would enable building societies to solicit additional investment. The government agreed to this arrangement on condition that 25 percent of funds raised through PUPS would be allocated for low-income housing loans.

The money raised through PUPS was thus allocated in two ways: 75 percent toward individual low-income mortgages and 25 percent toward the NHF (World Bank 1989a). After much negotiation, the building societies finally agreed to provide project beneficiaries with sufficient mortgage loans to cover the purchase cost of a plot (generally averaging about Z$600) plus the cost of constructing a minimum shelter of one room with an ablution facility. Plots were to be allocated according to the established procedures of the individual towns to those on their waiting lists. There was an upper income limit of Z$400/month, as it was expected that the incomes of 70 percent of the beneficiaries would be below Z$200/month, with income from lodging taken into account. The term of the loan would be twenty-five years with a variable interest rate (ibid.). Beverly Building Society, mentioned earlier, was the first to participate in this scheme. However, the government's introduction of tax-free PUPS soon attracted the CABS and Founders Building Society (FBS) to the program.

Urban II

On May 8, 1989, the Bank presented Urban II to its board of directors for approval, to run concurrently with Urban I. Urban II was the second major urban and regional initiative in Zimbabwe, and a relatively large project costing a total of US$580 million, to which the World Bank contributed US$80 million; a host of other international development sources contributed the remainder. (See Table 5.3 for the financing plan of the project.) Approximately 70 percent of the project's resources were dedicated to housing and housing-related infrastructure (Table 5.4).

Urban II was intended to build on its predecessor and co-runner, Urban I (discussed above), which the Bank regarded as successful. As Urban II's task manager, James Hicks, recalls, Urban I "laid a foundation for a housing program with a broader base outside the public sector."[29] The purpose of Urban II was to continue this trend and to "expand the role of private sector financial intermediation for housing from a pilot operation in four cities, to a nation-wide sector program."[30] The seven principal objectives of the project are summarized in Table 5.5.

Table 5.3 Financing structure for the Urban II urban and regional project

Organization	Contribution (US$m)
World Bank	80.0
West Germany	21.0
Swedish International Development Agency	3.0
Building societies (Zimbabwe)	242.0
Government (Zimbabwe)	234.0
Total	*580.0*

Source: World Bank (1989a: 2)

Implemented over a nine-year period beginning in 1990, Urban II comprised the following principal components (World Bank 1989a: 12–13):

* *Housing and related serviced residential land*: to provide support for a five-year capital investment plan for urban areas covering housing and housing-related serviced sub-sector loans to the government for residential infrastructure development. Funds were disbursed through the NHF for specific projects. House construction costs and the purchase of the serviced plots would be financed by long-term mortgages to be provided by the building societies directly to the individual beneficiaries.
* *Primary urban infrastructure*: to provide support for a five-year capital investment program for urban areas covering primary infrastructure (including maintenance requirements).
* *Regional development program*: to introduce a regional development strategy under which two pilot programs would be implemented. The first would provide technical assistance to prepare a strategic investment framework for the development of urban services infrastructure in the

Table 5.4 Components of the Urban II project

Components	Cost (US$m)
Housing infrastructure	79.3
Housing	248.3
Community facilities	15.1
Electricity	20.4
Infrastructure	70.9
Urban service and maintenance	42.5
Regional development program	10.6
Institutional development	9.1
Physical contingencies	24.1
Price contingencies	59.7
Total	*580.0*

Source: World Bank (1989a: iii).

Table 5.5 Objectives of the Urban II project

Provide support to urban centers in supplying additional urban services and housing that will be required as a result of the projected doubling of the urban population in seven to ten years

Strengthen the capacity of central and local government institutions to determine the most cost-effective investments in the urban sector and establish the most efficient scheduling of these investments

Ensure the continued financial self-sufficiency of the urban local authorities, including the maintenance of appropriate cost recovery and revenue generation capabilities

Protect the capital assets of urban local authorities through improved maintenance-related investments

Strengthen the human resource capacity of local authorities to enable them to efficiently deliver, operate and maintain urban services on a sustained basis.

Maximize the role of non-governmental private sector investors in housing, particularly building societies, as a means of relieving the financial burden on the government

Assist with the design and implementation of regional development programs focused on secondary cities, small towns, and rural centers as alternatives to the major cities, especially emphasizing employment creation in the urban areas to absorb the growing labor force

Based on World Bank (1989a: 12).

secondary towns along the Bulawayo/Harare/Mutare railway corridor. Emphasis would be placed on generating employment opportunities and promoting small enterprise development; the latter would assist with promoting the growth of selected small towns and rural centers in two provinces, Mashonaland East and Manicaland. This would include technical assistance to support, on a pilot basis, preparation of selected district development investment plans in infrastructure and services related to employment support measures as a means of expanding and diversifying non-farm activities.

*   *Strengthening institutional capacity* through a manpower development program that would include technical assistance and training to central government line ministries and local authorities. The program would address local financial management as well as urban service and maintenance requirements.

In contrast to Urban I, which was based in four cities, Urban II conformed to the Bank's own concurrently evolving urban agenda (discussed in Chapter 4) by emphasizing the need for nationally based urban initiatives. Allowing for the participation of all major towns meeting certain eligibility criteria,[31] the project aimed to effect "delivery of infrastructure, housing and related services of towns containing virtually all the urban population of Zimbabwe" (World Bank 1989a: 57). Like Urban I, Urban II included the provision of serviced residential land for low- and lower-middle-income households in all of the eligible cities.

Appropriate design standards were an issue in this project as well. The government agreed to reduce the minimum plot size from 300 m² to 200 m² for this project, which, the Bank then estimated, would enable the program to reach over 500,000 people (including occupants and lodgers). The program also intended to provide community facilities in the form of primary schools, clinics, market areas, and community centers.

As in Urban I, the building societies provided the necessary finance for the purchase of plots as well as the construction of dwelling units. Their role was greatly expanded in Urban II since they provided housing finance in the amount of US$240 million, for both plot acquisition and house construction, as well as loans averaging around Z$10,000. The Bank regarded the involvement of the building societies in both Urban I and Urban II as an important strategy in reducing the financial burden on the public sector[32] and envisioned that building society participation would eventually "restructure the financing of low-cost housing so that government funds are replaced by private sector resources."[33] As in Urban I, building societies were awarded grants to update their computing equipment in order to manage the anticipated increase in the volume of mortgages under the project.

Taken together, the World Bank's and the Zimbabwean government's monetary contributions to Urban II were to be directed in several ways (World Bank 1989a: 26) (Figure 5.4). First, some funds were to be allocated to the General Loan Development Fund (GDLF), the Ministry of Local Government, Rural and Urban Development (MLGRUD), the NHF, and the Ministry of Public Construction and National Housing (MPCNH). Money was then to be lent out from the GDLF through local authorities for the improvement and maintenance of traffic and urban services, as well as primary infrastructure. From the NHF, money was to proceed through local authorities to funds for housing-related on-site infrastructure. Second, Bank and government contributions to the building societies for the purpose of purchasing computer equipment were to come through a loan to the Ministry of Finance, Economic Planning and Development (MFEPD). Finally, World Bank and government contributions to the Zimbabwe Electricity Supply Authority (ZESA) were to be disbursed through the MFEPD.

Through its urban programs in Zimbabwe, the Bank encouraged private sector involvement in low-income housing. The Zimbabwean government also embraced the idea of involving the previously spurned building societies as it experienced great difficulty addressing the housing crisis immediately after independence. While building societies were initially reluctant to get involved, expediency and incentives persuaded them to participate in the program.

The World Bank hailed their entry as a success because the initial fear of potentially high default rates proved to be unfounded. Defaults were rare up

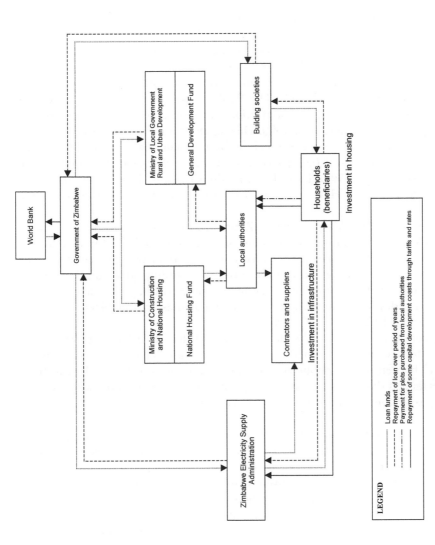

Figure 5.4 Flow of funds for Urban II. Source: World Bank (1984b: 19).

to the mid-1990s; in fact, a Bank review found that default rates were actually higher for upper-income groups (World Bank 1994b). When allottees did have difficulty meeting their monthly payments, high demand enabled them to sell their properties at a profit. Furthermore, building societies were able to do the same with repossessions.

By the closure of Urban I in 1994, some 18,000 residential plots and community facilities had been made available for low-income residents. The housing component of Urban I in Harare was initially concentrated in the black townships of Glen Norah, Sunningdale, and Budiriro. Urban II, in contrast, was implemented country-wide and involved over twenty-one cities and towns in Zimbabwe. The Bank's report of the project stated that some 30,000 stands for low- and middle-income housing were built before the project's completion in 1999. During the course of the project, 29,328 mortgage applications were received by the building societies. Of these, 22,432 were approved (compared with a target of 24,000) for the value of approximately Z$570 million (US$16 million). Approximately 11,200 stands were completed in Harare alone for the duration of Urban II (World Bank 2000b).

The World Bank's projects in Zimbabwe led to some gradual, limited improvements in housing conditions and access to facilities for some of the urban poor. The projects were also able to extend private sector finance to low-income families on a limited basis. According to a building society spokesperson, these programs not only supported sites-and-services schemes, but also developed a market for existing homes in the black townships (Beresford 1992).

From these findings, the Bank concluded that its urban projects forged an effective link between the public and private sectors and reduced the fiscal burden on the public sector. In its evaluations of Urban II, the Bank proclaimed that the successes and benefits of these new sector-wide projects surpassed the project-by-project approach of the 1970s:

> The project provides an excellent example of a case where the traditional public sector role in housing is reduced from that of total provision to the more limited one of servicing residential land, leaving the financial and actual contraction of the dwelling units to the private sector. The projects demonstrate how low-income households can have access to finance by providing an enabling environment for the housing sector.
>
> (World Bank 2000b: 8)

Thus, Urban I and II were celebrated by the Bank as solutions to the problem of low-income housing finance in Zimbabwe. The merit of this assertion is evaluated in the next section.

Critique of the World Bank's urban programs in Zimbabwe

The Bank claimed that Urban I and II created an alternative, enabling, and sustainable structure for low-income housing finance in Zimbabwe. This evaluation, however, does not stand up to empirical scrutiny. In addition to logistical problems encountered during the projects' implementation, there is some incongruity between the Bank's own evaluations and the projects' actual impact. This section aims to show that Bank programs did not influence the broader policy climate in a manner that improved the lives of the urban poor, contrary to its own initial expectations and final claims. I shall highlight four problem areas in my critique. First, the programs did not reach the poorest segments of the urban population, who were excluded because of eligibility criteria and lack of finance. Second, absentee landlordism and raiding by high-income groups prevented the poor from reaping the benefits intended for them. Third, while there were some successes, the programs failed to address the scope of the low-income housing problem in urban Zimbabwe. Fourth, the program failed to improve the lives of the urban poor in light of the deteriorating political and economic climate of the mid to late 1990s. Therefore I shall argue, contrary to the Bank's claims, that Urban I and Urban II did not significantly impact the problem of low-income housing in Zimbabwe.

Eligibility and benefits

Aiming to address the housing needs of those with incomes below Z$400, Urban I and Urban II initially estimated that at least 70 percent of the beneficiaries would belong to the sub-Z$200 income bracket. In practice, it was difficult to meet the needs of this income bracket because one of the participating building societies, CABS, was willing to consider only applicants with incomes over Z$250/month. The poorest households, therefore, did not benefit from the projects: in Harare, just before the projects were implemented, approximately 21 percent of the households on the waiting list for low-income housing earned less than Z$150/month, and overall, 50 percent of households on the list had incomes under Z$250/month and did not meet the cut-off for CABS loans (Harare 1989).

In Urban I, the World Bank estimated that 56 percent of low-income households would be able to afford two-roomed core houses, while 25 percent would be able to afford four-roomed core houses (an income of Z$380/month was required for the latter). The affordability estimate for four-roomed core houses was based on the assumption that three of those rooms would be let out for approximately Z$126/month. The building societies, however, refused to take rental income into consideration when allocating mortgages, which meant that households with incomes below Z$380 could not qualify for a mortgage to build a four-roomed house. In Urban II, the building societies did consider rent from two bedrooms as part of income, which made the four-room house affordable to households with incomes of at least Z$270/month. In Budiriro township, Harare, the mean income level stated by beneficiaries on their applications was Z$308/month. Only 1 percent of the beneficiaries earned Z$200/month or less, and 8 percent earned Z$400/month or more (Harare 1991). However, escalating costs in the late 1980s caused the income ceiling to be increased to Z$450/month in 1988, and to Z$550/month in 1989, which effectively excluded nearly three-quarters of the urban poor on the waiting list (Harare 1989).

From these numbers alone, it is clear that Urban I and Urban II did not reach the poorest segments of the urban population, despite their aims to do so. Thus, most of the urban poor could not meet the eligibility criteria for building society loans. Harare's waiting list stood at around 60,000 households in the mid-1990s and was estimated to be growing at a rate of 900 additions per month. However, the City Council was able to service only some 1,500 plots per year (Harare 1995).

From the Bank's perspective, the primary objective of Urban II program was to encourage private sector financing for low-income housing in the long term:

> The private sector will provide all financing for low-income housing. Broader-based investment in building societies will occur through the creation of several new mechanisms to raise long-term housing finance. The result is that the societies' low-cost mortgage portfolio will expand and the government's long-term portfolio will correspondingly diminish.[34]

However, as seen earlier, private sector financing meant that a substantial number of the urban poor could not gain access to funds to meet their housing needs. In this respect, Rakodi (1995) has suggested that lower-income households that are not eligible for private sector finance, and thus unable to participate in the Bank's program, should be funded by the government on a subsidized basis, or have access to alternative means of finance.

Table 5.6, presenting the economic profile of waiting lists for key urban centers in 1995, indicates that the majority of those on the waiting list could not afford even the most basic kind of housing offered through the World Bank/building society schemes.

The problem of speculation

The serviced plots for low-income housing in Zimbabwe were allocated on a freehold basis with no restrictions on resale. Rakodi (1997) observes that one of the problems encountered with such schemes in other countries was allottees' realization that considerable profit could be made by selling out and realizing the full market value of the property. In Zimbabwe, however,

Table 5.6 Economic profiles of applicants on waiting lists in various urban areas

| Place | Income (per annum) | | | | |
| | *<Z$2,000* | | | *<Z$1000* | |
	Number	%		Number	%
Harare	62,606	88		48,840	68
Bulawayo	22,757	94		19,617	81
Chitungwiza	32,388	80		N/A	
Mutare	14,099	93		11,395	75
Gweru	8,730	97		6,660	74
Masvingo	6,402	96		5,178	78

Source: Government of Zimbabwe (1996).

these types of sales were generally limited by the lack of alternative housing, as well as the relatively greater value of home ownership during difficult times. Therefore, the severe housing crisis and the shortage of plots for houses for medium-income purchasers attracted buyers who had, on average, higher incomes than those for whom the plots were intended. According to Rakodi and Withers (1995), average prices of houses for sale on the market in high-density areas increased gradually between 1986 and 1988, but then skyrocketed by 103 percent between 1989 and 1990. The increase during the early 1990s was around 63 percent. In April 1991, the advertised prices of homes on the market ranged from Z$30,000 to Z$175,000, averaging around Z$90,000, whereas the purchase price of a serviced plot, for a four-roomed house, including construction, was only Z$18,000 during the same period (Table 5.7).

Rising prices effectively shut the poor out of the housing market, even from the houses that were intended for low-income families. A resident of the satellite township of Chitungwiza describes the problems that resulted:

> The fact of the matter remains that Chitungwiza's housing shortage is not only desperate, but has also reached the point where it has become pathetic, especially when one takes into account the fact that on average up to 30 people live in a single residential unit. It is even sadder when one considers that the majority of Chitungwiza's residents are so-called lodgers who occupy dwellings of varying sizes at rent to die rates [*sic*] . . . In fact, the exorbitant rents that the majority of the people in Chitungwiza are paying, the anguish, and the frustration that has become so evident means that the time bomb cannot be far from exploding.[35]

The pervasiveness of absentee landlordism in Harare's high-density suburbs was highlighted in a speech by B. Mayo, a member of parliament:

> Some of the problems that affect the provision of adequate housing in the local authorities are caused by the rich who buy off houses in high

Table 5.7 Average advertised price for houses in high-density areas

Average house price (Z$)	1986	1987	1988	1989	1990	1991
Tafara/Mabvuku	17,188	–	–	25,500	28,000	39,952
Mbare	–	–	–	–	46,500	63,769
Mufakose	23,500	–	23,107	–	43,182	44,615
Dzivarasekwa	24,053	19,750	34,143	26,333	48,000	71,667
Glen Norah	23,879	25,000	18,292	25,300	44,833	94,961
Glen View	22,028	21,000	28,141	29,000	68,539	100,473
Highfield	24,067	19,000	30,125	40,000	66,666	175,228
Kambuzuma	22,417	32,833	26,154	33,143	75,000	131,667
Kuwadzana	20,350	12,500	23,596	27,615	62,143	104,167
Warren Park	25,000	20,600	23,450	37,500	89,000	117,083
Budiriro	–	17,000	40,000	–	50,333	105,909
Total	22,583	23,841	25,724	30,068	61,133	99,789

Source: based on Rakodi and Withers (1995: 264).

density areas and use people fraudulently whom they exploit and make them pay rent [*sic*].[36]

Absentee landlordism and speculation in the developing world is related to the problem of petty accumulation, according to Iliffe (1983). When other avenues of accumulation are closed, property becomes a means of accumulating wealth. As a result, the poor are burdened by higher prices and rents. Iliffe observes that petty accumulation is a hindrance to dynamic capitalist development in Africa:

[Due to] the easy availability in modern Africa of alternative investment opportunities which were more profitable but less productive than manufacturing industry . . . Urban property has generally been a more secure investment than trade or industry, easier to finance by means of loans from banks or the state, and simply more profitable. It was reckoned in the 1960s that capital invested in house-building in the low-income area of Mathare Valley in Nairobi could be wholly recovered in rent in 18 months. For Nairobi's African bourgeoisie in recent years – in the past for Europeans and Indians – property ownership has been described accurately as the "express lift to prosperity."

(ibid.: 68–9)

Rising urban land and housing prices also pushed housing schemes for lower-income households further into the periphery of the city. Hence, low-income households had to pay more to live in areas that lacked basic services and were far away from jobs (Patel 1988). While the higher prices tapered off by the mid-1990s, low-income households in urban Zimbabwe still faced great difficulty in securing adequate finance for housing because of deteriorating political and economic conditions.

The scope of the housing crisis

While the Bank intended its programs, especially Urban II, to provide an alternative system of financing for low-income housing, the impact of these programs did not match the extent of the problem. In the first three years of lending for low-cost housing, just under 14,000 mortgages were issued, compared with the nearly 16,000 mortgages issued for high-cost housing, although the need for the former was much greater (Government of Zimbabwe 1991). In addition to the building societies' fiscal conservatism in issuing loans, local authorities' delays in servicing the plots also contributed to the problem and prevented many poor people from participating.

Table 5.8 shows the average costs for various sites-and-services schemes in Harare in the mid-1990s. A basic 150-m² plot cost around Z$6,500. With a twenty-five-year mortgage at 15 percent, this type of serviced plot was estimated to require a monthly payment of Z$173 (Z$63 mortgage + Z$110 owner charges) (Kamete 1999). Even if a household was able to allocate 30 percent of its income to meeting housing costs, a monthly income of Z$577 would be required to purchase the basic plot. Figures like these suggest that the World Bank's programs fell far short of meeting the needs of low-income households and failed to confront the market as a barrier to housing delivery. A USAID report acknowledged this inadequacy:

> On the positive side, government policy has encouraged home ownership through the sale of rental housing owned by central and local governments and the private sector to the occupants of these homes. This policy is furthered by the active implementation of a housing guarantee scheme

Table 5.8 An analysis of affordability in the Kuwadzana Low-Income Housing Project (all costs in Z$)

	Housing levels			
Payments	Plot only	Plot + wet core	Plot + two rooms	Plot + three rooms
Cost of land	6,500	18,500	27,500	33,500
Deposit (25%)	1,625	4,625	6,875	8,375
Principal	4,875	13,875	20,625	25,125
Monthly repayments at 15%	63	179	266	324
Total home ownership charges	110	110	110	110
Total monthly cost (A+B)	173	289	376	434
Minimum income required	577	963	1,253	1,447

Source: based on Palmer Associates (1995).

which greatly facilitates the purchase of housing for the growing urban middle income group. [Since independence], however, new construction has been quite low indicating that the housing stock has deteriorated and not improved.

(USAID 1985: 45)

Thus, the major beneficiaries of the Bank's programs tended to be lower-middle to middle-income families, and not the urban poor, as the Bank had intended. Currently, access to low-income housing in Zimbabwe still looks bleak for this group, with unemployment rates soaring higher than 50 percent in cities and median household incomes remaining around Z$310 (Palmer Associates 1995; Kamete 2001). Furthermore, international development agencies such as the World Bank have generally withdrawn recently from low-income housing finance in Zimbabwe, as have the building societies, because of a deteriorating economic and political climate (discussed in the next section).

In 1999, the National Housing List Register indicated that the housing backlog in Zimbabwe had reached the 1 million mark. The list, which registers households and not individuals, indicates that one-third of all Zimbabweans are lodgers and lack even a bare minimum of housing.[37] While Zimbabwe does not have the large-scale squatter settlements that characterize many cities of the developing world, squatting is hidden by lodging and rental accommodation.[38] The fact that lodging is fast emerging as a "solution" to the lack of low-income housing is clearly visible in Harare. The 1992 census (Government of Zimbabwe 1992) indicated that 66 percent of the urban population in the city were either living with extended family members, lodging (renting a room or two), sleeping on the street, or squatting. This problem is likely to increase annually given the current low rate of housing delivery at present. While detailed reports of the 2002 census are not yet available, preliminary results indicate that the trend is on the rise.[39]

In the high-density black suburb of Chitungwiza, I met many families with four or five children living in one small room in 1993.[40] In 1995, it was reported that it was common for as many as five families to reside in one unit in the high-density suburbs of Mbare, Highveld, and Glen View in Harare.[41] A study conducted by the city of Harare in 1992–3 (Harare 1994) revealed that in the high-density area of Highveld there were about eight lodger families per homestead with a family size of three to five, which meant that in a 300-m^2 unit there were approximately thirty-six people, including the owners. In Mbare, Highveld, and Epworth, many lodgers share an outside water tap and communal toilets, which also function as bathrooms. Highveld lodgers have complained that more than twenty families frequently use the same water tap and toilet/bathroom facilities.[42] In Epworth, water sources are between 0.5 and 1.5 km from the homestead (Butcher 1986), which is a great inconvenience. Evelyn, a resident, told me:

> My brother and I walk to the tap every morning to fetch water. It is far
> away and we have to wait in a queue when we get there. We pump the
> water and carry it in those buckets there. I am often late to work because
> of the water.[43]

Congestion also poses a number of problems for the urban low-income
population. For many of lodgers that I spoke with, the single room they
rented was their entire living space. It was where they ate, slept, cooked, and
socialized. Privacy was virtually non-existent and strangers shared intimate
spaces, which posed numerous problems for family and other relationships.
Elizabeth, a 34-year-old domestic worker in Mount Pleasant and a lodger in
Chitungwiza, shared her frustration with the lack of space:

> It is very crowded here. Sometimes we are crawling over each other in
> this small room. My children sleep here on this side under the table. I
> sleep on the cot there because I have arthritis. I can't get up from the
> floor. All of our things are there under the cot and on the table. Those
> people sleep on that side. There are four of them . . . The children always
> fight with my children. I don't say anything because I have nowhere to
> go. I am a widow with three children. I have to stay here.[44]

Not surprisingly, such crowded conditions also resulted in numerous health
problems.

In sum, the World Bank's programs may have increased access for a few
low-income residents, but they have only marginally affected the overall
critical housing shortage in Zimbabwe, which is primarily due to insufficient
availability of finance for housing development. The NHF's resources have
been exhausted and have not been replenished.[45] To complicate matters
further, the World Bank suspended aid to Zimbabwe as of November 2000
because of political instability. However, even with international assistance,
as the previous discussion showed, the majority of households on the waiting
lists did not qualify for finance under existing eligibility criteria. At present,
the housing crisis is compounded by the international development agencies'
policies of fiscal management and austerity, as well as the imprudent choices
of the domestic elite in Zimbabwe. This issue is addressed in the next
section.

Some political and economic concerns

The previous discussion highlighted specific organizational problems in Urban I and II, especially their eligibility criteria and their impact. The impact
of Urban I and II has to be seen in relation to broader political and economic
processes and some key issues must be considered: (i) the relationship of the
projects to the national development agenda after independence; (ii) the ef-

fects of the IMF/World Bank structural adjustment programs; and (iii) the impact of domestic political choices on the poor's ability to meet their housing needs. As mentioned earlier, the Zimbabwean government opted for a pragmatic approach to development following independence, abandoning its rhetoric of fundamentally restructuring the old order. Concomitantly, neither USAID nor World Bank programs attempted to transform the colonial basis of economic and spatial relations in Zimbabwe. The first USAID program in Kuwadzana, for example, did not acknowledge the issue of the excessive distance (11 km) from employment and shopping offered in the city. World Bank programs similarly accepted the geography of the former regime. For example, the World Bank's urban sector report did not address the need to incorporate Chitungwiza into the Harare municipal area in a meaningful way. Nor were there any proposals to integrate low-income urban residents into the former white residential areas, other than through the market. For instance, two residences already existed on most plots in the low-density suburbs: the main house and servants' quarters. Schlyter (1990) notes that servants could have bought their houses on the same terms as tenants who bought the city council houses. Such solutions, however, were too radical for the new "socialist" government and the international development agencies.

In sum, very little transformation of the existing urban form has taken place since the demise of white minority rule in Zimbabwe. The Bank, as pointed out earlier in the chapter, viewed the problem as one of "extreme dualism," and recommended an extension of the market-based system into the black townships as a solution (World Bank 1985b). However, the differences and unequal relationships between black high-density townships and former white residential areas in Zimbabwe are better captured by the idea of uneven capitalist development (Smith 1984; Bond 1998). Whites used their political power to distribute economic resources in their own favor; high-income, low-density suburbs resulted as segregationist policies marginalized the African population. Thus, independent Zimbabwe inherited a housing system that embodied the economic, political, social, and spatial asymmetry inherent in settler colonial society. With the exception of upper-class blacks who could afford to move into the low-density suburbs (Pickard-Cambridge 1988; Cumming 1990), the colonial urban space economy was left intact by the government and the World Bank.

Consequently, a number of the urban poor's hardships were unaddressed by both government and World Bank initiatives. The white regime had deliberately located the high-density black townships on the urban periphery in order to separate the racial groups, as noted earlier (previous chapter). As a result, low-income residents incurred high transport costs commuting to work. According to Musiyazviriyo (1992), the total average commuting time was approximately four hours, one way, for low-income residents from high-density black townships to the center of town. This meant that most workers spent at least sixteen hours a day away from home, which left very little time for family life, or even rest for that matter. Matthew, a sixty-year-old gardener from Chitungwiza working in Mount Pleasant, commutes to work daily.

> I have to be here at 8:00 but in order to get here on time, I have to be on the bus at 6:00. I get to the bus-stop at 5:00 to make sure that I don't miss the bus because if I have to wait for another one, I will be late to work . . . I leave for Chitungwiza at 6:00 and reach there at 9:00. I am very, very tired by the end of the day.[46]

Dengura (1995) shows that transport costs for the urban poor have also soared in recent years. In the early 1980s, low-income commuters were spending about 8 percent of their incomes on transportation to Harare, whereas by the mid-1990s they were spending between 22 and 45 percent of their already meager incomes on transport alone. Table 5.9 lists the costs of transportation for low-income commuters to Harare in 1995. More recent figures indicate that transport costs for the average manual worker from Chitungwiza increased to 50 percent of monthly income.[47]

Continuity with the urban forms of the previous regime reflects the unwillingness of both the domestic elites and international development agencies to address the fundamental causes of urban poverty in Zimbabwe. Although Zimbabwe's liberation movement was premised on transforming the existing order, that aim was virtually abandoned by the 1990s. A left-wing ZANU member of parliament, Lazarus Nzareybani, lamented in 1989 that

> The socialist agenda has been adjourned indefinitely. You don't talk about socialism in a party that is led by people who own large tracts of land and employ a lot of cheap labor. When the freedom fighters were fighting in the bush they were fighting not to disturb the system, but to dismantle it. And what are we seeing now? Leaders are busy implementing those things which we were fighting against.
>
> (cited in Bond 2000: 96)

The social welfarist concern of the early independence years diminished under the influence of the World Bank and IMF and under the changing

Table 5.9 Bus routes and weekly fares in Harare, Zimbabwe

Route taken	Bus weekly fare (Z$)	Percent of monthly income
Seke to Glen Norah	22.00	22
Budirio to Central Harare	18.00	24
Highveld to Central Harare	12.00	29
Chitungwiza to Central Harare	22.50	45

Source: Dengura (1995: 45).

domestic priorities of the ruling elite (Astrow 1983; Herbst 1989; Sylvester 1991; Weiss 1994).

The second reason for the limited gains of Urban I and Urban II was the impact of structural adjustment programs on the domestic political economy in Zimbabwe. In March 1991, the Zimbabwean government developed and implemented what came to be known as a "home-grown" Economic Structural Adjustment program (ESAP) (Stoneman 1993: 89). ESAP mirrored the programs of the World Bank and the IMF in that it included devaluation of currency, export promotion, trade liberalization, privatization of government enterprises and para-statals, and reduction in expenditure in the social service sectors such as education, health, and housing.

Assessing whether "external capitulation or domestic reform" caused the Zimbabwean government to embrace market-based reforms and privatization, Dashwood (1996: 29) argues that, while Zimbabwe had been under considerable pressure from the World Bank and IMF to liberalize its economy since 1982, the "initiative for reforms came from within Zimbabwe." However, once the reforms were implemented, the influence of the World Bank and IMF grew considerably. By the mid-1980s, the Zimbabwean ruling class had reached a consensus that economic reforms, along the lines of those advocated by the World Bank and IMF, were necessary to promote faster economic growth and better domestic response to changing international conditions. This led to a convergence of views between the ruling classes and international development agencies on market-based reforms and privatization. The establishment of a World Bank mission office in Zimbabwe in 1985 put the Bank in a good position to influence the views of the ruling elites. The Bank quickly got to work and began to produce a series of policy studies outlining recommendations for changes in domestic policy (World Bank 1987, 1989a, 1990).

ESAP was far from successful. Its targets for growth and development were missed by huge margins and real wages fell by about 30 percent following its implementation in the early 1990s (Stoneman 1999). Bond (1993) convincingly demonstrates that ESAP was misguided from its inception. The exclusive focus on exporting primary products to an increasingly hostile international economy, which was bogged down in recession, did not generate internal economic growth in Zimbabwe. (See Table 5.10 for the decline of the Zimbabwean dollar.) Furthermore, Zimbabwe had very little control over the international prices of the primary commodities, which plunged to historic lows during the 1990s. Consequently, formal sector employment declined and public services in health, education, and housing were cut deeply, thereby worsening the plight of the urban poor, and making it extremely difficult for them to meet their basic needs (Gibbon 1995; Tevera 1995).

The deteriorating macroeconomic climate, induced in part by the structural adjustment programs, eroded the limited gains of Urban I and II. The Bank's technocratic emphasis on reducing and/or eliminating

Table 5.10 US$ to Z$1 conversion table

Year	US dollar (US$)
2005	0.00001732
2004	0.00023240
2003	0.00143385
2002	0.01816993
2001	0.02
2000	0.03
1999	0.03
1994	0.14
1991	0.35
1989	0.46
1987	0.59
1985	0.64
1983	1.00
1981	1.43

public sector spending on social programs failed to generate alternatives for addressing the needs of those on the economic margins of Zimbabwean society. In examining the impact of the Bank's structural adjustment programs and private sector initiatives, ul Haq (1998) notes that, although the need for developing countries to trim their budgets is not disputed, the downsizing was always done in a manner that adversely affects the poor. He particularly pointed out how military expenditures actually increased while social expenditures decreased in many countries undergoing World Bank adjustment programs, but the former are seldom targeted for reduction:

> The bitter controversy over the need for structural adjustment programs often missed the real point. Of course budgets needed to be balanced. But the real issue was, what expenditures were being reduced? A very disturbing picture emerges. The poor countries slash their education and health expenditures, while increasing their expenditures on the military, with the *World Bank and IMF watching silently from the sidelines.*
>
> (ibid.: 15, emphasis added)

Similar patterns may be observed in Zimbabwe. While the World Bank and IMF supported austerity measures that eroded whatever minimal public sector support there existed for the poor, the ruling classes in Zimbabwe continued to enjoy extravagant lifestyles and often used state structures for their own advancement. The domestic policy choices of these elites constitute the third factor limiting the impact of Urban I and II, as discussed below.

The consequences of ESAP's liberalization, privatization, and market reform were borne not by the ruling elite, but by the poor. Following the success of the national liberation movement, many whites left the country, abandoning farms and businesses, but also bureaucratic and managerial posts in the cities. The luxurious residences of the departing whites in low-density suburbs were available for cheap purchase immediately after Zimbabwe's independence. As a result, a small proportion of the black population experienced upward mobility in both the public and private sectors, but such gains hardly touched the colonial legacy of inequality.

Robert Mugabe's and ZANU's coming into power in 1980 was hailed as a victory for the oppressed masses, and socioeconomic inequalities in the country were expected to be eradicated soon. However, as the ruling elite consolidated its position, it increasingly abandoned its earlier radical rhetoric, identified with international and national financial interests, and alienated the urban and rural poor. At the political level, this elite accommodationism was enabled by the unity of accord between the two major political parties, the Zimbabwean African Union – Patriotic Front (ZANU-PF) and ZAPU, leading to a merger of the two parties in 1989. Soon, the state and the ruling ZANU-PF party became indistinguishable. Dashwood (1996) observes that the merger allowed the elites from both parties to reap the economic benefits and privileges of their powerful positions within the ruling structures, at the expense of the rural and urban poor.

The new ruling elites have all but abandoned the goals of the national liberation struggle, plundering the state's resources for their own gains. Corruption is now a major impediment to development in Zimbabwe. Stoneman and Cliffe (1989) note that rural tracts of land intended for poor peasants ended up in the hands of members of government. In the absence of government structures of accountability to address abuses of state power and privilege, the popular press and emerging opposition movements were faced with the task of exposing the wealth-accumulating tendencies of the ruling elite. There were a number of reports of mistreatment and poor working conditions on farms owned by the elites. For example, it was reported that laborers on the farm of Dr. Kombo Moyana (a governor of the Reserve of Zimbabwe in 1992) lived in squalid and overcrowded conditions.[48] The former commander of the Zimbabwean National Army, Tapfumanei Solomon Mujuru, built up a business empire worth millions of US dollars in the name of his brother Misheck Mujuru.[49] Individuals also took advantage of their associations with members of the ruling classes in order to solicit government contracts. For example, a Z$1.2 billion tender to build a new airport in Harare and a Z$250 million tender to supply it were awarded to Leo Mugabe, the president's nephew. The glaring transparency of this case led a normally pliant parliament to turn down the contracts.[50]

In March 1997, a major national scandal erupted when it was discovered that Z$450 billion was "missing" from the War Victims' Compensation Fund, which was set aside to compensate ex-combatants for injuries suffered during the Liberation War. Mugabe appointed a judicial commission of inquiry chaired by Justice Godfrey Chidyausika, which then revealed that senior officials in the political and military wings of government, including the late First Lady's brother, had appropriated the fund.[51]

The country is currently crippled by the War Veterans' protest and subsequent occupation of white-owned farms. The powerful Zimbabwe National Liberation War Veterans Association, under the leadership of Chenjerai Hunzvi, staged mass demonstrations from June through July 1997, meeting with Mugabe and finally compelling him to agree to a lump sum payment of Z$50,000 to all ex-combatants who fought in the national liberation struggle, plus a Z$2,000 monthly allowance. Since many of the ex-combatants populate the military and police forces, Mugabe agreed to these demands without consulting his cabinet or parliament, putting additional stress on the country's already weak economy. Furthermore, Mugabe announced that 1,480 mostly white-owned farms would be seized and 20 percent of the land would be distributed to the war veterans.[52] On November 14, 1997, also known as "Black Friday," Mugabe's unbudgeted Z$4 billion settlement with the war veterans resulted in the collapse of the Zimbabwean dollar, which fell by 75 percent in just a few hours. Interest rates were increased by 6 percent in the course of the next month, as were sales and petrol taxes, in order to help cover the costs of this scheme (Brickhill 1999).

Private sector finance for low-income housing, an integral part of the Bank's housing programs, evaporated in this deteriorating political and economic climate. According to Colleen Butcher, building societies were still willing to finance low-income housing, but it was not viable for them to do so in such an environment. With interest rates skyrocketing, building societies were investing their deposits in money markets, not issuing mortgages.[53] Given the absence of a public sector housing program in Zimbabwe due to the factors discussed above, the urban poor were once again forced to rely on their own meager resources to house themselves.

The poor experienced additional hardships because of the structural adjustment programs and price hikes in the aftermath of Black Friday. Their massive and violent demonstration over rising food prices in Harare in January 1998 left nine dead and hundreds injured and caused over Z$70 million in damage. The contempt of the ruling elites for the plight of the poor was demonstrated by the fact that, during the same week, the government announced that it had spent some Z$60 million on fifty new Mercedes-Benz automobiles for twenty-six cabinet ministers and two vice-presidents. Additionally, instead of committing resources to provide for the basic needs of Zimbabwe's own lower classes, the Mugabe government committed troops to a foreign war in the Democratic Republic of Congo at a cost of nearly US$1 million a day.[54]

With respect to low-income housing in particular, about US$3 million from the NHF and the National Housing Guarantee Fund was "borrowed" by Mugabe's wife, Grace, in order to build herself a thirty-two-room house with three servant cottages in a rich suburb of Harare in 1998. Nicknamed "Graceland," the house is hardly used because the First Lady "changed her mind about living so far away from the city centre."[55] Chinawa, a lodger in a high-density township who ekes out a living selling vegetables nearby, demanded to know how "the government [can] go about putting up taxes and asking people to be patient when a house like this is built and left empty?"[56] The government's only reply was that "homelessness has become a part of everyday life"[57] (Government of Zimbabwe 1996).

The crisis facing the urban poor was exacerbated in May 2005, during the Zimbabwean winter, when government security forces launched Operation Murambatsvina (which means "clean out the trash" in Shona), a massive slum and squatter removal campaign in Zimbabwe's major cities. In the once flourishing city of Victoria Falls some 30,000 people were evicted from their informal settlements, and in the capital city Harare entire squatter settlements were burnt down (*Independent*, June 12, 2005). President Mugabe declared that the mass eviction "Operation Murambatsvina was needed to restore sanity to Zimbabwe's cities"(reported by the BBC, June 17, 2005). In an address to the central committee of the ruling ZANU-PF party, Mugabe characterized the demolitions as an important part of the "urban renewal" of Zimbabwe:

> Our cities and towns had become havens for illicit and criminal practices and activities which just could not be allowed to go on. From the mess should emerge new businesses, new traders, new practices and a whole new and salubrious urban environment. That is our vision.[58]

The Minister for Local Government, Ignatius Chombo, also characterized the squatter eradication campaign in utopian language, claiming, "This is the dawn of a new era. To set up something nice, you first have to remove the litter, and this is why the police are acting this way" (*Mail and Guardian*, June 9, 2005). In this manner, the homes of 200,000 urban poor were destroyed. The United Nations estimates that some 700,000 people have been displaced as a consequence of this policy and 2.4 million more were adversely effected by the government's actions (United Nations 2005a).

The Mugabe government's rationale for this policy was that it would restore law and order, curb the chaotic growth of informal settlements, and strengthen the formal economy. The degenerating conditions in Zimbabwe's cities during the last ten years were also caused by the neo-liberal policies implemented by Mugabe under the guidance and instruction of the World Bank. The ESAP program, mentioned above, contributed to severe job losses in the urban areas. Consequently, the workers had to generate their own incomes in the informal sector at the government's own urging. In fact, the government pressured local authorities to relax standards to accommodate informal economic activities. However, in a puzzling change in policy, the

Mugabe regime became obsessed with "the illegality" of the informal economy in May 2005.

It is now apparent that a major factor prompting Mugabe's assault on the urban poor was his desire to crush the Mass Democratic Movement (MDM) opposition party, because the primarily urban-based MDM threatened the ZANU-PF stranglehold on power. Welshman Ncube, secretary general of the MDM, characterized operation Murambatsvina as a "harassment campaign against urban voters."[59] In the March 2005 elections, Mugabe's ruling ZANU-PF party lost the cities to the MDM. Many ZANU members' homes and informal businesses were destroyed in the government's retributive campaign, which not only aimed to curb dissent, but also attempted to gain control of the informal economy that had operated outside of the government's grip during the last few years. Elliot Manyika, ZANU-PF's national commissar, said that "the economy needed to account for informal business and order needed to be restored in urban areas."[60] During its raids, the government discovered large caches of foreign currency in the urban homes of informal traders. The campaign against informal markets is expected to enable the Mugabe government, which faces a dire shortage of foreign currency, to get a better grip of foreign currency transactions.[61]

However, instead of restoring law and order, this campaign has only deepened the country's political and economic crisis. Inflation is around 144 percent, the unemployment rate is estimated to be approximately 70 percent, and some 3 million people are facing the prospect of starvation (*Business Day*, September 12, 2005). Current World Bank president Paul Wolfowitz has characterized the situation as "a tragedy."[62] However, such a statement is somewhat disingenuous from the head of an institution that bears at least partial responsibility for the misguided policy choices of the Mugabe regime.[63] Structural adjustment contributed to structural collapse in Zimbabwe.

Conclusion

Zimbabwe's critical shortage of housing for the low-income urban poor ranks next to unemployment as the most serious problem confronting the country in the post-independence period. From the mid-1980s, the government courted the assistance of the World Bank in plotting a strategy to address the housing problem. The Bank's entry into urban development in Zimbabwe coincided with shifts in its own urban lending strategy from a project-based to sector-based approach. In fact, an internal Bank memorandum outlining a lending strategy for southern Africa noted that the Bank "will undertake new initiatives in the shelter sector only when we can effectively address major sectoral issues."[64] Hicks recalls that Urban I and II were designed with the intention of impacting Zimbabwe's entire urban sector:

The Bank was concerned that its earlier urban work in squatter upgrading and sites-and-services did not have the desired sector-wide
impact. We were very conscious of the Bank's lending philosophy in the
1980s, that projects should have a sector-wide impact. While the projects
in Zimbabwe dealt with housing for the poor, we came at it from the
angle of the delivery system and sought to engage the private sector in
low-income housing delivery on a sustainable basis.[65]

When the Bank presented Urban II for board approval, it argued that:

The proposed project would replace public with private sector finance for
low-income housing. The project will increase the supply of affordable
low-income housing on a full-cost recovery basis and secure long-term
and sustainable finance from the private sector for low-income housing
... It will fundamentally reform the housing delivery structure for low-
income housing in Zimbabwe.

(World Bank 1989b: 3–4)

In keeping with its development philosophy since the 1980s of reducing
public expenditure on social programs and promoting privatization, the
Bank's urban projects in Zimbabwe incorporated building society finance
for low-income households. While the projects did enable some low-income
urban families to gain access to credit, they certainly did not provide an
"alternative and sustainable structure" for low-income housing finance or an
"enabling" environment for low-income housing delivery, as the Bank claims.
This is because, I argue, Urban I and II were coordinated with Zimbabwe's
own existing ESAP program, which emphasized fiscal austerity, reductions in
public social expenditures, and privatization. The net effect of ESAP, which
was constructed in consultation with the Bank, was the erosion of the poor's
ability to meet their own basic needs. A series of domestic crises, combined
with economic and political backlashes to the ESAPs, have resulted in the
virtual withdrawal of private sector institutions from low-income housing
finance and, with it, the Bank's much touted "solution" to low-income housing
finance and delivery. The poor are now assisted by a few philanthropic NGOs[66]
or struggling to house themselves with their own meager resources. Of the
total of 2.3 million housing units in the country, piped water is available inside
only 324,000 units, with water available outside for 536,000 units. Electricity
is connected to only 450,000 homes.[67] As a consequence of this shortage and
neglect, Nyoni observes, "Zimbabwean cities are in rapid decline. Lodging
and backyard houses are proliferating, and low-cost housing areas are ill-
supplied, if at all, with basic services and amenities."[68] The World Bank's
market-oriented approach to low-income urban housing has hardly touched
the 7.2 million people who live below the poverty line in Zimbabwe. The
multilateral development agencies' fundamentalist faith in the "magic of the

market" and fiscal austerity has not only failed to offer any viable strategy for dealing with growing inequality, but has actually exacerbated it.

As this and earlier chapters have shown, although international forces have impacted development in Zimbabwe, the role of the national state in the process cannot be ignored. Fainstein (2001: 295) observes that "national governments may not be able to affect the global economy, but they can shield citizens from the most pernicious effects of that economy." In the Zimbabwean case, instead of protecting the majority of its citizens [the urban and rural poor] from the adverse effects of the global economy, the ruling elite have not only ignored the fact that another world existed in the African townships; they have actually worsened the plight of the poor through their ill-conceived policies and parasitism upon the public.

Chapter 6

Globalization, neo-liberalism, and the politics of the World Bank's current urban agenda

Globalization and neo-liberalism have altered the nature and function of the state in both developing and developed countries. These twin pressures reinforced the World Bank's conservative development philosophy in the 1990s. Three neo-liberal economic premises have undergirded Bank policy since the late 1980s (Bruno 1994: 2):

- Attainment of sustained average per capita growth is a necessary condition for sustained reduction in poverty.
- Implementation of an adjustment package of policy reform is a necessary condition for sustained per capita growth.
- Fiscal and monetary restraint is a necessary condition for adjustment.

Acting on these premises, the World Bank pressured governments in developing countries to privatize or eliminate social welfare programs and promote fiscal conservatism until state policy began to mirror the behavior of the private sector. State involvement with social concerns became increasingly entrepreneurial, more committed to creating favorable investment environments for private capital than to equity or social justice.

This chapter identifies some key problems with the free market fundamentalism evangelized by the Bank so fervently in the 1980s and early 1990s. Next, I describe some recent protest against Bank policies from civic groups and various NGOs. In light of their campaign, the Bank tried to reorient its agenda under James Wolfensohn. This section describes how social welfare concerns, largely ignored during the 1980s, were back on the Bank's agenda. However, as the concluding section of the chapter shows, reform efforts were jettisoned after they created tensions within the institution and among Washington power brokers who were committed to the neo-liberal agenda. Current urban initiatives at the Bank continue to draw upon and promote neo-liberal principles.

The World Bank's free market fundamentalism

Privatization of public sector enterprises became one of the pillars of the Bank's urban agenda during the 1980s and early 1990s, as shown in Chapter

4. During the 1970s, under McNamara, the Bank's urban work represented a distinct departure from traditional Bank practices. Chapter 2 tracked the Bank's foray into poverty alleviation, especially through partial sites-and-services schemes and squatter upgrading in cities. The urban agenda of the 1980s, in contrast, increasingly conformed to the emerging neo-liberal paradigm and articulated with a policy trend identified by Merrifield as "lean urbanization":

> As the Dow got fat and bullish, not only has American industry gotten lean but cities have gotten lean as well, and a sort of lean urbanization pervades and dramatizes the current American urban condition, dramatizes people's contemporary experience of urbanism. Just as Wall Street has rewarded corporate shakedowns, job eliminations, downsizing and rightsizing, it rewards lean cities, too – or at least rewards landed property and investors within lean cities, those personifications of abstract space, those who can really make space pay.
>
> (Merrifield 2002: 161)

The World Bank and IMF have ensured that "lean urbanization" has become the prevailing global ideology in much of the developing world during the 1980s and 1990s. In accordance with the Bank's de-emphasis of the role of the state, the new urban agenda championed privatization and targeted the dismantling of public sector housing, arguing that it was inefficient and ineffective in addressing the housing needs of the urban poor. Under structural adjustment, shelter was seen as part of a social sector that did not directly contribute to increased export earnings for the given country. Government involvement in housing, it was argued, discouraged private initiative, encouraged corruption, and increased bureaucratic inefficiency. As noted in the previous chapter, it was within this policy context that the World Bank entered Zimbabwe. However, transferring responsibility for low-income housing in Zimbabwe from the state to households and the private sector failed to address the problem, as the previous chapter illustrated.

As public sector support was reduced and subsidies were eliminated, the costs of low-income housing were increasingly borne by low-income residents themselves under lean urbanization. The Bank's current emphasis on full cost recovery discriminates against low-income beneficiaries of Bank projects by placing upon them a financial burden seldom borne by the middle and upper-income occupants of private-sector housing developments. Cost recovery has generally been confined to the level of individual projects, instead of being spread over a wider sectoral basis. The capital costs of all off-plot infrastructure and off-site works are charged directly to the project beneficiaries instead of being spread across the whole urban population, for instance, through the general/local tax structures, as usually done with the extension of infrastructure. The failure to explore creative alternatives to cost recovery has meant that the urban poor are able to afford housing only through a reduction in construction standards and service provision. As a result, plot sizes are too small to allow for real quality of life; in fact, they have adversely affected public health and safety in these communities. For instance, some plots financed by the Bank's Urban Development Project in Madras, India, were only 25 m^2 (Wakely 1999). While Bank-funded plots in Zimbabwe were not this small, the reduction in standards was nevertheless the only response to escalating costs that was ever pursued seriously.

During the 1980s and 1990s, the Bank promoted full cost recovery and market-driven urban development with such evangelistic zeal that questioning its new orthodoxy has been likened by some to "whistling into a typhoon" (Jones and Ward 1995: 69). Even though the Bank has denied any ideological underpinning to its thinking by defending its policies as "the only way to remedy past urban distortions" (World Bank 1994a: 28), shifts in Bank policies nevertheless ought to be interpreted as responses to particular historical moments, as I have argued throughout this book, rather than as outcomes of politically neutral, technocratic evaluations of project and program efficacy, as Bank officials have maintained. Samir Amin has observed that:

> [World Bank] experts love to brag of their "political neutrality." They pride themselves on the hidden defect of many economists desirous of being technocrats, capable of mentally shaping a "good development policy," "scientific," "devoid of any ideological prejudice." But this kind of exercise has the supreme virtue of avoiding the real options facing currently existing societies. The truncated and superficial image of reality characteristic of this genre under discussion must of necessity lead to false conclusions.
>
> (Amin 1990: 35)

Amin's observation is evident in the Bank's dichotomization of the private and public sectors, which portrays the former as dynamic and energetic, and the latter as moribund, lethargic, and lacking initiative. Such a separation obscures the ongoing relationship between the two spheres. Some point out that the vitality of the private sector is dependent on public sector finance and retired public officials often work as consultants to the private sector, and suggest a degree of interdependence and cooperation between the two sectors (Sanyal 1986). Yet, what is often touted as a public–private partnership really consists of a one-way gradient of resources flowing out of the public sphere into the private. Such a relationship can hardly be characterized as symbiotic. In fact, now, given the rescaling of the state and the push toward privatization under the neo-liberal orthodoxy, the relationship is actually parasitic upon the public.

The Bank frequently ignores this reality when it exalts the private sector's role in development. For example, Taiwan, South Korea, Hong Kong, and Singapore are often hailed as paragons of market-led growth with booming economies. However, an adequate supply of housing by the state played an integral role in the development of these societies, according to Castells *et al.* (1990). "While conservative ideologues such as Milton Friedman and . . . all the Thatcherites of the world" see these countries as the "showcase[s] for the ideology of economic *laissez-faire*," they ignore the important role the much reviled public sector played in these economies:

> Government supported housing, health, education, transportation, and subsidies of foodstuffs and basic daily consumption items have been crucial elements in ensuring a proper production and reproduction of labor, in making labor cheaper without lowering its quality, in providing a safety net that has enabled an entrepreneurial population to take risks by investing and creating businesses and in providing the basis for social stability since the early 1970s.
>
> (ibid.: 4)

Advocates of the privatization of housing delivery frequently ignore some of these historical realities. When questioned about the failure of the prevailing neo-liberal orthodoxy in meeting the urban poor's needs, Bank policy-makers are adamant that their policy measures were not applied rigorously enough, and refuse to look outside their limited paradigm. As Milder (1996: 162) observes, "The Bank still believes that countries have two choices: the Bank's way or failure."

It has become difficult for critics of Bank policy to penetrate what has come to be known as the "Washington Consensus Doctrine." The term was coined by John Williamson to summarize the policies that "Official Washington" (World Bank, IMF, US Treasury) regarded as appropriate and necessary for growth in the developing world. The principles of the Doctrine are (Williamson 1990):

- fiscal discipline;
- a redirection of public expenditure priorities toward fields with high economic returns and the potential to improve income distribution, such as primary health care, primary education, and infrastructure;
- tax reform (to lower marginal tax rates and broaden the tax base);
- interest rate liberalization;
- a competitive exchange rate;
- trade liberalization;
- privatization;
- deregulation.

While Williamson himself is supportive of some of these reforms, and coined the term to describe and summarize them (Williamson 2004), its current usage has been transformed from his original formulation and appropriated by critics. "Washington Consensus" is now commonly understood to mean the policy prescriptions of the neo-liberal orthodox establishment (Kanbur 1999).

Contrary to the Bank's claim that its policies actually enhance growth and improve the welfare of the developing world, I have argued in this book that they have, in fact, exacerbated inter- and intra-state polarization. Inequality has climbed over the past three decades, with the poorest 20 percent of the world's population watching their share of global income decline from 2.3 percent to 1.4 percent (Castells 1998). The incomes of the richest 20 per cent, on the other hand, rose from 70 to 85 per cent during the same period (ibid.). According to the UNDP, rising poverty and inequality have been characteristic features of the 1980s and 1990s:

> Since 1980, there has been a dramatic surge in economic growth in some 15 countries, bringing rapidly rising incomes to many of their 1.5 billion people, more than a quarter of the world's population. Over much of this period, however, economic decline or stagnation has affected 100 countries, reducing the incomes of 1.6 billion people, again more than a quarter of the world's population. In 70 of these countries average incomes are less than they were in 1980 – and in 43 countries less than they were in 1970. [Furthermore], during 1970–85 global GNP increased by 40 per cent, yet the number of poor increased by 17 per cent. While 200 million people saw their per capita incomes fall during 1965–1980, more than one billion people did in 1980–1993.
>
> (UNDP 1996: 1–2)

Despite the claims made for it, the Washington Consensus Doctrine mostly failed to make economies viable and undermined economic prospects for millions of the world's poorest people.

As a result of this glaring failure, the Bank faced a serious crisis of legitimacy by the early to mid-1990s as a number of organizations mobilized against its policies. Even by the late 1980s it was becoming increasingly evident that structural adjustment programs and conservative fiscal policies were generating far more problems than they were resolving (United Nations 1997; SAPRIN 2004). Their negative repercussions for the urban poor took the form of higher prices, reduced wages, increased unemployment, and cuts in social services. Urban poverty increased under the structural adjustment programs (SIDA 1995).

The World Bank versus civil society

As the Bank approached its fiftieth anniversary in 1994, a number of NGOs and other civil society groups launched a series of sustained protests against the policies of the World Bank and IMF. The Rainforest Action Network took out advertisements in the *New York Times* attacking the policies of the Bretton Woods institutions. The advertisement's headline read "How to Borrow Millions and end up Homeless" and featured a picture of a poor woman begging with a bowl.[1] The protests escalated in 1994 as the annual meetings of the World Bank were proceeding in Madrid. Bank president, Lewis Preston, was confronted with numerous demonstrations as he arrived. Protestors built a refugee camp on the median strip of the main highway from the airport to downtown Madrid in order to draw the Bank's attention to the plight of the poor under structural adjustment. Other nonviolent but nevertheless potent acts of civil disobedience were carried out in the heart of the city to embarrass the World Bank. As Preston was speaking during a press conference, a protestor managed to get onto the stage and unfurl a banner right behind Preston that proclaimed "World Bank Murderer" in bold letters. Preston's image against that backdrop was duly featured on the front pages of all the major newspapers.[2] That event was obviously a fiasco for the Bank, but the worst was still to come.

As Lewis Preston (not the most captivating public speaker) was delivering his keynote address at the annual meeting, fake dollar bills bearing the inscription "World Bankenstein" began to flutter down from the ceiling of the Madrid conference center. Two young activists had climbed onto the roof of the facility and released the dollar bills. While the security personnel were baffled, Preston himself was not fully aware of the commotion, owing to his poor eyesight, and labored on with his speech. If the previous event was embarrassing, this one was a public relations disaster for the World Bank. Official Washington decided that the Bank desperately needed a new face. As the problems with structural adjustment became more difficult to deny and public protest began to mount, Bill Clinton appointed James Wolfensohn in 1995 to become the ninth president of the World Bank.

As shown in the previous chapters, the Bank has a long history of successfully adapting to changing political contexts and surviving as an institution by preserving and promoting its own relevance. Under Wolfensohn's leadership, the Bank once again launched a series of initiatives to transform the image of the institution at the rhetorical level, back to that of the anti-poverty crusader. Immediately after assuming office, Wolfensohn embarked on a tour of Africa, just as McNamara had, to gain first-hand knowledge of the Bank's programs on that continent. There, he aggressively pursued the NGO community that had protested Bank policy so vigorously. For example, when Tony Burdon, Oxfam's representative in Uganda, told Wolfensohn at a meeting that he was "sorry that we don't have time to meet" Wolfensohn rescheduled his meeting with the Ugandan Minister of Finance in order to get onto Burdon's calendar and meet other NGO leaders in Kampala (Mallaby 2004: 109).

Through various outward gestures, Wolfensohn attempted to show that the Bank was distancing itself from the Washington Consensus Doctrine. Upon his arrival at the Bank, he proclaimed his vision for the institution with a new sign at the entrance to the World Bank headquarters in Washington DC that read, "Our Dream is a Poverty-Free World." In World Bank press releases and his own statements, Wolfensohn and some senior Bank staff conceded to critics and acknowledged the limitations of the market-driven agenda in addressing the multifaceted needs of the developing world. Next, Wolfensohn focused on the criticisms of the NGOs, which seemed to be enjoying much popularity and legitimacy at the grassroots level. He, too, embraced the idea of debt forgiveness for the poorest countries and pressured senior management to cancel Nepal's Arun III dam project, conceding to the demands of environmental groups. He invited NGO representatives to play a role in policy formulation and analysis. Wolfenson's courting of the NGOs soon began to yield some gains. Whereas NGO protests at the 1994 annual meetings were a serious blow to the Bank's image, the 1995 annual meetings in Washington DC were a public relations victory. Wolfensohn had succeeded in persuading Oxfam, one of the leading NGO critics of the Bank, to share the platform with him in discussing debt relief (Mallaby 2004).

Internally, echoing McNamara, Wolfensohn tried to broaden the Bank's focus and reintroduce equity concerns into the Bank's agenda, by speaking in the language of the critical NGOs:

> There was so much emphasis on economic transition that the social consequences were forgotten ... All the focus was on numbers ... It became clear to me that there was a need to have a different analysis ... There were issues of education and health, women's rights, water and power which are not addressed.[3]

In order to meet this challenge and broaden the Bank's mandate, Wolfensohn ordered the Comprehensive Development Framework (CDF),[4] which stated that, in addition to promoting growth and free market economies, Bank programs ought to address social safety nets, establish efficient legal systems, and support a host of other social concerns. With the CDF, Wolfensohn urged the institution to move away from free market fundamentalism and board the "sustainable development" bandwagon, so that the Bank might begin to define what that term ought to mean.

> We need a *second* framework, one that deals with the progress in structural reforms necessary for long-term growth, one that includes the human and social accounting, that deals with the environment, that deals with the status of women, rural development, indigenous people, progress in infrastructure, and so on.
>
> And so in our discussions at the Bank, we have developed and are experimenting with a new approach. One that is not imposed by us on our clients but developed by them with our help. An approach that would move us "beyond projects," to think instead much more rigorously about

what is required for sustainable development in its broadest sense.

The framework would call for policies that foster inclusion – education for all, especially women and girls. Health care. Social protection for the unemployed, elderly, and people with disabilities. Early childhood development. Mother and child clinics that will teach health care ... [The] framework would describe the public services and infrastructure necessary for communications and transport. Rural and trunk roads. Policies for livable cities and growing urban areas, so that problems can be addressed with urgency – not in twenty-five years when they become overwhelming. And alongside an urban strategy, a program for rural development that provides not only agricultural services, but capacity for marketing and for financing and for the transfer of knowledge and experience.

(Wolfensohn 1999, various pages)

With a multipronged strategy that promoted the combined use of markets as well as social opportunities, Wolfensohn promised to make the dream of a poverty-free world come true. In an effort to demonstrate this commitment, he created a managing directorship for social issues and recruited Mamphela Ramphele, former vice-chancellor of the University of Cape Town, South Africa, and former black consciousness activist, to the position.

In addition to external criticism, Wolfensohn was responding to internal calls for the reform of Bank policy, especially from the newly appointed senior vice-president and chief economist at the Bank, Joseph Stiglitz. Stiglitz criticized the tenets of the Washington Consensus, calling its doctrine "misguided" and too narrow in focus:

We have to broaden the objectives of development to include other goals, such as sustainable development, egalitarian development and democratic development. An important part of development today is seeking complementary strategies that advance these goals.

(Stiglitz 1998a: 1)

In his opinion, proponents of privatization "overestimate [its] benefits and underestimate [its] cost" in developing societies because "left to itself, the market will tend to underprovide human capital." Privatization particularly poses a problem for poorer families who do not have access to the market for shelter, health, education, and other needs. Stiglitz argued that in these instances governments should make some provision for the needs of marginal groups. Stiglitz also felt that the Bank should display "greater humility" in dealing with the developing world and acknowledge that "[it] does not have all the answers" (Stiglitz 1998a: 7–13).

The Washington Consensus Doctrine confused means with ends: it took privatization and trade liberalization as ends in themselves, and not as means for more sustainable, equitable, and democratic growth. In an interview, Stiglitz recommended that three critical issues needed to be addressed if the Bank was to move away from the problematic Doctrine (Stiglitz 2000a):

1   the Bank ought to broaden its definition of development to include democratic, equitable, and sustainable development, and not focus exclusively on economic growth;
2   the Bank needs to change its underlying economic philosophy – there is very little benefit, Stiglitz argued, in pushing down inflation at the cost of higher unemployment and social distress; and, finally,
3   the Bank needs to change the tone of the development dialogue, especially its "colonial mentality" toward the developing world.

Stiglitz (ibid.) argued that multilateral development agencies act as if they alone know what is best for the developing world, and prevent effective dialogue on what might be done.

In an effort to address these concerns, James Wolfensohn joined Klaus Töpfer, the executive director of the United Nations Centre for Human Settlements (UNCHS) (Habitat), in launching an initiative called Cities Alliance in 1999 and invited Nelson Mandela to be its patron (Hildebrand 2001). Cities Alliance aimed to introduce equity concerns into the Bank's urban programs in a manner reminiscent of the McNamara era. For example, a key component of Cities Alliance is Cities Without Slums, an action plan calling for squatter upgrading schemes to improve the living conditions of 100 million people over the next twenty years (ibid.).

Persistent neo-liberalism

In time, such initiatives caused disagreement at the Bank between the orthodox economists and supporters of the "new" comprehensive development agenda, such as Stiglitz. This conflict lay behind the political circumstances leading to the resignations of Joseph Stiglitz and his associate, Ravi Kanbur.[5] Stiglitz had appointed Kanbur in 1997 to head a team to author the 2000 world development report, titled *Attacking Poverty*. The original draft of this report[6] became quite controversial over the next three years because it argued for a "more flexible, free market doctrine," and questioned the existing policies of the World Bank for its minimal positive impact on socioeconomic inequality in development (Kanbur 2000: 4). Trade reforms, SALs, etc. were acknowledged as benefiting upper-income groups while hurting the bottom 40 percent of the population. The report also launched a scathing critique of the World Bank and the IMF's management of the Asian crisis, especially policies requiring Korea, Thailand, and Indonesia to slash their budgets and enact various World Bank/IMF-prescribed reforms in order to qualify for assistance. The report even admitted that these measures "instantly unraveled a generation of progress against poverty." Citing the Indonesian example, the report pointed out that "school dropout rates for the poorest group of children surged from 1.3 per cent to 7.5 per cent in 1998" as a consequence of World Bank and IMF measures, whereas Malaysia, which had refused IMF aid and implemented various controls on investment flows against the advice of the World Bank and IMF, "had a shallower recession and faster recovery than its neighbors." Finally, while the report recognized the importance of economic growth, it argued that "empowerment, security and opportunity" should complement strategies that emphasize growth. "Empowerment," in this case, referred to how state institutions ought to address the needs of the poor and remove social barriers. "Security" meant making the poor less vulnerable to economic fluctuations, natural disasters and economic crises, while "opportunity" emphasized the need for growth (ibid.: 11–13).

Storm clouds gathered as Official Washington received the draft of the report put forth by the reformers at the Bank, led by Joseph Stiglitz. Treasury Secretary Lawrence Summers[7] and Stanley Fischer, chief economist at the IMF, immediately objected to the report, especially the section that attributed some of East Asia's economic problems to "the rapid opening up of markets to short-term capital flows" (ibid.:14). The Treasury Department was infuriated by the report's praise of Malaysian capital controls as wise restrictive measures that ought to be used as standard tools for promoting the health of Third World markets (Bullard 2000; Sivaramakrishnan 2000). The orthodoxy at the World Bank and the IMF followed suit and attacked the report's emphasis on empowerment, particularly its advocacy of enabling the poor to form "networks, cooperatives, trade unions and the like so that they may voice their political and economic concerns" (Kanbur 2000: 16). Official Washington also had probems with the section on security because it warned that economic reform promoting free trade should not take place in the absence of security measures to cushion market failure.

Opponents circulated the draft to a number of sympathetic intellectuals in order to boost their case against the report. T.N. Srinivasan, a Yale University economist who had collaborated with Anne Krueger (the conservative former chief economist of the Bank under Clausen) on a number of research projects, launched the following tirade:

> Security, opportunity and empowerment could at best be termed as diagnostics and at worst as three symptoms of the disease or syndrome of poverty, but they certainly do not provide an analytical engine.
>
> (cited in Wade 2001: 132)

Angus Deaton from Princeton University echoed Srinivasan and was joined by other macroeconomists at the Bank and IMF who argued that the 2000 world development report compromised market development. In response to the mounting chorus, Wolfensohn ordered Stiglitz to make the report conform more closely to *laissez-faire* principles. Citing his objection to such pressures to "edit" the document, Ravi Kanbur resigned from the project in June 2000.

Wolfensohn, who had frequently used Stiglitz's ideas to demonstrate the Bank's openness to criticism against the Washington Consensus Doctrine, was now at a loss. The Treasury Department continued to exert great pressure on the Bank to renounce Stiglitz and his reformers. In fact, US Treasury Secretary Lawrence Summers eventually made Wolfensohn's own reappointment for a second term contingent on Stiglitz's resignation as chief economist and senior vice-president at the Bank.[8] Presented with such a choice, Wolfensohn asked Stiglitz to resign in August 2000, but retained him temporarily as his own senior policy advisor. However, Stiglitz was soon forced to leave the Bank altogether because he had added "fuel" to the "fire" caused by the report draft: He had published an article in the *New Republic* earlier that year against the Washington Consensus and the manner in which the Bank had handled the Asian crisis. Stiglitz was critical of the alliance between the Bank and the IMF in jointly funding structural adjustment loans in general, but of the Fund's economists in particular:

> the older men who staff the Fund – and they are overwhelmingly older men – act as if they are shouldering Kipling's white man's burden. IMF experts believe they are brighter, more educated, and less politically motivated than the economists in the countries they visit. [The IMF] economists lack extensive experience in a country; they are more likely to have first hand knowledge of its five-star hotels than of villages that dot its countryside.
>
> (Stiglitz 2000b: 57)

Stiglitz also argued elsewhere that the Bank would be more successful in fulfilling its mission of reducing poverty and implementing Wolfensohn's comprehensive development agenda if it divorced itself from the IMF altogether:

> Many developing countries need assistance because they are poor. Structural adjustment suggests that they are out of the kilter, that they need a nose job. My point is that they're poor and need more money to be less poor. If the IMF is restrained from long-term lending, and the Bank and IMF no longer present a united front, then the Bank will be freer to move ahead in this direction.[9]

Stiglitz's departure and the controversy over the 2000 world development report cast great doubt on the Bank's supposed "shift" away from the Washington Consensus Doctrine. During the Wolfensohn presidency, the Bank tried to move away from the orthodox economic models it favored during the 1980s and 1990s. At least in the public rhetoric of its highest officials, the

Bank went to great lengths to argue that growth is not enough and that its programs should be sensitive to the needs of the poor. "Poverty reduction" became part of the Bank's official vocabulary again, as in the McNamara years, and a few of its programmatic initiatives, like Cities Alliance, are even explicitly directed at the poor. Nevertheless, the models of the 1980s and early 1990s continue to inform Bank policy and have left a strong imprint on the Bank's current thinking. For example, its call for a reduction in the role of the state in development and its undying, fundamentalist faith in the market remain major cornerstones of its philosophy, leaving one to wonder how much has really changed at the Bank: structural adjustment-type loans still constitute 63 percent of the World Bank's lending (Mallaby 2004), in spite of the administrative relabeling of projects and scattered references to ecology, poverty reduction, and civil society. Behind these semantic incorporations and makeovers under Wolfensohn, however, lies the conservative orientation that the Bank has retained since its inception.

In fact, the Bank's current urban initiatives are replete with neo-liberal policy recommendations. In 2000, the Bank launched a major program called Capital Markets at the Sub-National Level in order to "increase and develop the financial capacity of local governments and . . . assist local governments in accessing capital markets financing" (World Bank 2000d). According to Sven Sandström (2000), managing director of the Bank, current economic realities necessitate that local governments of developing countries raise funds from international capital markets. The purpose of the Capital Markets program, according to Sandström (ibid.) is to assist LDCs in identifying their financial needs and negotiating the complexities of international private sector borrowing. He writes:

> the reality is that many sub-national governments do not have a track record in place to make them attractive borrowers. Despite the fact that they have come a long way during the past decade, significant challenges remain. Faster corporatization and privatization of municipal services are critical; so are better legal and regulatory frameworks. Further progress in institutional capacity building is necessary, as is clarity of central government rules and guidelines. And governance reforms, as well as increased transparency at the local government level, are required. These are the building blocks needed for development of improved financing structures, improved communication between the private and public sectors, and better service delivery. They are also necessary conditions to expand market financing for sub-national governments, and to minimize the risk to overall fiscal stability that undisciplined sub-national borrowing creates.
>
> (ibid.: 6)

One of the major components of the Capital Markets initiative is the issuing of municipal bonds to raise funds for municipalities. The Bank's present Urban Division argues that the success of the US municipal bond market suggests that such a program in the developing world can assist local governments in financing critically needed infrastructure without borrowing from the central government by enabling them to access capital markets directly. In fact, the Bank is encouraging municipalities to proceed with the program without central government guarantees in order to ensure effective decentralization, good management, and creditworthiness. In addition, the Bank argues that the overall cost of such projects can be less than that of other types of finance, such as public–private partnerships and borrowing from multilateral development agencies like itself.

In 2000, the Bank sponsored a major international conference to promote the Capital Markets initiative, which I attended. City executives and other local government officials from developing countries, representatives of credit rating agencies, and investors from around the world gathered at the New York Hilton to deliberate on how they might participate in and support Capital Markets. In promoting the merits of the program, Bank officials and private sector speakers repeatedly cited the conclusions of the *New York Times* journalist Thomas Friedman's (1999) book, *The Lexus and the Olive Tree*. Friedman's widely celebrated book almost seemed to have been written for the occasion. His central metaphor of the "Golden Straitjacket" was repeatedly invoked by almost all the plenary speakers to highlight the futility of deviating from free market, capitalist development for Third World countries. In his book, Friedman divined that, although people are unhappy with the "Darwinian brutality of free-market capitalism," there is no ideological alternative at present:

> When it comes to the question of which system today is the most effective at generating rising standards of living, the historical debate is over. The answer is free market capitalism ... When your country recognizes this fact, when it recognizes the rules of the free market in today's global economy, and decides to abide by them, it puts on what I call "the Golden Straitjacket." The Golden Straitjacket is the defining political-economic garment of this globalization era. The Cold War had the Mao suit, the Nehru jacket, the Russian Fur. Globalization has only the Golden Straightjacket. If your country has not been fitted for one, it will soon be.
>
> (ibid.: 85–7)

In line with the Washington Consensus Doctrine, Friedman argues that in order to fit into the Golden Straitjacket a country must adopt or at least be seen as moving toward ensuring that the private sector is the main engine of growth in that country. The country ought to shrink the size of its state bureaucracy, eliminate or lower tariffs in imported goods, and accept other policies based on the Washington Consensus Doctrine. Once all these policy pieces are in place, Friedman writes, they will be "stitched" together to produce a Golden Straitjacket for the country. The idea is universally applicable, according to Friedman, although "unfortunately, this Golden Straitjacket is pretty much 'one size fits all'." Nations which wear the "Golden Straightjacket" shall experience two things, Friedman prophesied: "your economy grows, your politics shrink" (ibid.: 87). This idea was sold with great zeal at the Capital Markets conference as the basis for the Bank's program of assisting local governments to gain access to capital markets.

It was interesting to observe that a number of delegates from developing countries were very curious about Friedman's book after hearing its continuous praise at the conference. I was seated with the Zimbabwean delegation, who could not wait to purchase the book after it was presented at three or four sessions as the bible of the contemporary global economy. The Zimbabweans were very excited about learning how their country could be fitted into a "Golden Straitjacket." During a break between sessions, four members of the Zimbabwean delegation asked me where they might purchase this "insightful, new book." Although I was not surprised by Bank officials' praise of the book, I nevertheless had to marvel at how they managed to sell such a one-sided view on development. However, what did surprise me, in a counterintuitive way, was the eagerness with which a group of delegates from a recently "socialist" Third World country imbibed that view.

No alternative viewpoints to the "Golden Straitjacket" were presented at the conference in spite of the Bank's rhetoric of "comprehensive development." There was no mention of Indian Nobel Laureate Amartya Sen's (1999) important work, *Development as Freedom*, which is critical of development ideologies that focus exclusively on the market. Sen's book gives serious consideration to issues of governance, equity, poverty, and gender, which were part of the Bank's official lexicon. Published contemporaneously with Friedman's book, Sen's work was based on a series of lectures that he had delivered as a Presidential Fellow at the World Bank in 1996 and 1997, when Wolfensohn was speaking of steering the Bank away from the Washington Consensus Doctrine. Despite their monumental importance, not even a faint whisper of Sen's ideas, nor their implications for urban development, was heard at the Capital Markets conference.

At present, the World Bank is energetically promoting its "new" urban agenda of facilitating local governments' access to international financial markets, seemingly oblivious or unaware of the potential for fiscal and political instability. While the Bank wants local governments to be economically self-sufficient, the capacity of local governments to meet all their capital requirements without some form of central government assistance is doubtful, as ably argued by numerous critics, including Stiglitz and Sen. For example, a problem arises if a loan is not guaranteed by the central government, raising the fundamental issue of where the debt lies. In other words, if the central government is not involved, is the loan to be regarded as part of the internal debt of a country or external to it? Who would redeem the debt if the central government is not obligated to do so? The local government? National sovereignty is at stake as the Bank tests these new ideas at the municipal/local level.

The World Bank's Urban Division is at the forefront of this rescaling effort, aggressively pushing the neo-liberal logic of globalization through its Capital Markets initiative. Its current urban agenda is influenced by an emerging school of thought known as "convergence theory," which, according to Michael Cohen (1996), is based on the idea that cities of the developing and developed worlds alike are plagued by a common set of problems, and require similar policy responses. While globalization is an undeniable force in the contemporary world economy, the notion of a borderless world, promoted by some theorists (Ohmae 1990; Prazniak and Dirlik 2001), is debatable. The idea that economic interests will become so strong that markets will replace politics ignores the political fact that structures and actors are rescaling to adapt to emerging realities and to remain relevant.

Chapter 7

Conclusion

Since its founding at the Bretton Woods conference, the World Bank has been a pivotal actor in development. As the largest multilateral development organization in the world, its role in the development process has been the subject of much deliberation and debate. This book has examined how the Bank's definition of the goals and methods of development have shifted over time and how those shifts have impacted the Bank's urban lending agenda. This work has attempted to demonstrate that the periodization of the Bank's development philosophy according to the geo-political context of the times is vital to explaining how and why the Bank embraced urban lending during the early 1970s as well as subsequent shifts in the Bank's urban agenda.

During the 1970s, the Bank's shelter policies advocated an active role for government in upgrading informal settlements and offering land tenure to squatters as well as providing other services to those areas. While the sites-and-services and squatter upgrading programs were not without problems (see Chapter 3), they constituted the few times that the Bank's urban programs ever addressed the needs of the urban poor directly. However, just when developing countries were revising their shelter policies, upgrading informal settlements, and providing serviced plots to the urban poor, the Bank's urban agenda changed, partly because of internal factors and partly in response to the changing political terrain in Washington. The emergent conservative critique of the Bank's urban program claimed that the sites-and-services schemes and squatter upgrading amounted to handouts and welfare programs. Since then, the World Bank has shifted from a project-based approach that targeted specific urban problems, to urban programs that focused on entire sectors, and finally to macroeconomic policy aimed at restructuring the state and promoting privatization. Table 7.1 depicts these policy shifts. Through an integrative policy approach, the Bank attempts to influence the overall course of development in a country.

In his 1968 Nairobi address, Robert McNamara challenged the international community to eradicate poverty by the end of the century. However, five years into the new millennium, urban poverty and inequality remain pressing problems. Twenty percent of the world's population (more

Table 7.1 Types of bank lending

Lending type	Description
Project lending	This is the classic Bank loan: coal plants, oil development, fisheries, forestry, and agriculture projects, dams, roads, education, population planning, water/sanitation and health projects, urban projects, housing, etc.
Sector lending	These loans govern an entire sector of a country's economy (energy, agriculture, industry, urban). A single loan carries conditions determining the policies and national priorities for the sector
Institutional lending	The World Bank lends in order to reorganize government institutions, orienting their policies toward free trade and open access for transnational corporations. Privatized utilities but also regulatory bodies in many countries are products of World Bank lending
Structural adjustment lending	Structural adjustment loans were nominally intended to relieve the debt crisis, convert domestic economic resources to production for export, and promote the penetration of transnational corporations into previously restricted economies. Most southern countries have now undergone structural adjustment – often called austerity programs by local people – under World Bank and IMF auspices. In the 1990s, these programs were extended to the former Soviet Union, the former socialist countries of Eastern Europe, and India as well.
Integration policy	The four types listed above are coordinated in practice, particularly since the Bank reorganized in 1987, to provide a much stronger "country focus." The Bank's goal is now to make sure that *all* its loans to a given country contribute to the achievement of adjustment policy objectives there

Source: George and Sabelli (1994: 16–19).

than 1 billion people) eke out a living on less than US$1 a day, with another 2 billion subsisting on US$1–2 a day (United Nations 2005b). The lack of adequate shelter is particularly challenging, according to a recent United Nations report. About one-sixth of the world's population live in slums, and that number could double by 2030 if serious attention is not paid to the lack of adequate shelter in the developing world (United Nations 2003). The report mentions that the worldwide number of slum dwellers increased by 36 percent in the 1990s to 923 million people. The United Nations report places the blame for the proliferation of slum settlements squarely at the doorstep of the World Bank and IMF:

> Much of the economic and political environment in which globalization has accelerated over the last twenty years has been instituted under the guiding hand of a major change in economic paradigm – that is, neo-liberalism. Globally these policies have re-established a rather similar

international regime to that which existed in the mercantilist period of the 19th century when economic booms and busts followed each other with monotonous regularity, when slums were at their worst in Western cities, and colonialism held global sway. Nationally neo-liberalism has found its major expression through Structural Adjustment Programmes (SAPs), which have tended to weaken the economic role of cities throughout most of the developing world and placed emphasis on agricultural exports, this working against the primary demographic direction moving all of the new workers to towns and cities. These policies, as much as anything else, have led to the rapid expansion of the informal sector in cities, in the face of shrinking formal employment opportunities.

(ibid.: 6)

As the state increasingly retreated from addressing social concerns and local governments were pressured to adopt conservative fiscal policies by the World Bank and IMF, NGOs increasingly began to fill the void. NGOs have become a powerful voice in international development practice, championing important causes related to poverty reduction, lack of adequate shelter, women's rights, and a host of other social concerns. Heralded as a "third sector" in the economy and characterized "as new agents with the capacity and commitment to make up for the shortcomings of the state and the market in reducing poverty" (Paul 1991: 1), NGOs are said to be an important constituent in a "bottom-up" approach to development that aims to involve people more directly in the practices of development.

At present, many NGOs are at the forefront of articulating issues of social justice and equity that have all but evaporated from debates on development. Organizations such as Amnesty International, Greenpeace, Human Rights Watch, Care International, Oxfam, and a host of other NGOs have engaged in praiseworthy campaigns to fight against human rights abuses and address the basic needs of people whose plight would otherwise have been ignored or suppressed by national elites and international multilateral development agencies. For example, corporate exploitation has been held in check by anti-sweatshop movements, while a number of NGOs have publicized Royal Dutch/Shell's disregard for the rights of the Ogoni people in Nigeria.

The NGO coalitions that fought for debt forgiveness won an important victory for the poor of the developing world, when the G8 nations decided to forgive the debt of the poorest countries. As a result, Paul Wolfowitz, current World Bank president and former architect of war (Iraq), has become the latest convert to the "debt forgiveness" crusade. Without sustained activism by NGO coalitions like Jubilee 2000, the IMF and the World Bank would not have found it politically necessary to embrace debt forgiveness. The massive peaceful, nonviolent demonstrations of 1999 that disrupted the fiftieth anniversary celebrations of the World Bank and the World Trade Organization in Seattle challenged the undemocratic policies, projects, and practices of these organizations. The 50 Years Is Enough Coalition, founded in 1994 to coincide with the Bretton Woods institutions' fiftieth anniversary celebrations, played an important role in raising public awareness of the adverse effects of SALs on the world's poor. The coalition's forceful campaign for the fundamental restructuring of the World Bank and IMF themselves resulted in a "reorientation" of the World Bank's focus in the mid-1990s under Wolfensohn (discussed in Chapter 6). Joseph Stiglitz and Mamphela Ramphele (former managing director of the World Bank) have acknowledged that pressure from social movements and NGOs has caused the Bank to change its policy direction and put poverty reduction at the forefront once again (Stiglitz 2002; Pithouse 2003).

The World Social Forum (WSF) identifies itself as an "open meeting place for reflective thinking, democratic debate of ideas, formulation of proposals, free exchange of experiences and inter-linking for effective action, by groups and movements of civil society that are opposed to neo-liberalism." Such movements are part of an emerging global civil society that provides a forum for the poor, who otherwise might not be heard. Many NGOs are motivated by deeply moral and political concern for social justice and equality. They are peopled by sincere, dedicated, and intelligent individuals with tireless energy and youthful optimism, whose vigilance and activism have enabled millions of ordinary people to go about their everyday lives in countries where their basic rights and freedoms are being scaled back by national states in consultation with the World Bank and IMF.

However, there is another side to the NGO story that bears mentioning here. In spite of the noble work done by many NGOs, these entities are not without their problems. Many NGOs have ties to fundamentalist religious movements and have become neo-liberal agents themselves, much like their larger and more powerful counterparts, the World Bank and IMF. Some major differences, possibly advantages, are that they reach the grassroots directly and do not seem to suffer from crises of legitimacy that plague the World Bank and IMF. The fact that NGOs are often able to bypass the state in bringing resources directly to their target groups has aided the poor in many countries but presents new problems of accountability. Many NGOs are funded by the very entities that incurred the wrath of the poor during the "bread riots" and other protests: international development agencies, Western governments, and corporations. Because they are ultimately accountable to those who finance them, NGOs are poised to deploy the neo-liberal agenda in unique ways on behalf of the historic bloc (discussed in Chapter 1) with which they are hegemonically articulated. The access that many NGOs have to the grassroots strategically positions them to give form to popular demands. If the World Bank is able to capture and dictate agendas through macroeconomic policy, NGOs shape political possibility at a different point of origin: the people themselves. Thus, it is not surprising that the Bank is courting NGOs; articulation with these actors is a further rescaling strategy of the glocalizing World Bank. As the new voices of the public, NGOs are in a position to strike strategic compromises with historic blocs. These compromises eventually emerge as "solutions" to problems in the absence of state policy or the presence of state repression.

In a manner that seemingly inverts the "top-down" approaches of other actors within a given historic bloc, NGOs direct consent upward, connecting it to compromise at other scales, thus completing a scaled channel of policy formulation and implementation. Grassroots access also enables NGOs to inflect and manage public discontent effectively by inviting dissenters to "participate" in "self-reliant" or "cooperative" efforts. This gives people the impression that something is being done but the effect is to contain popular rage at the grassroots and prevent its escalation toward the state and corporate interests. The fact that politics in many developing countries is articulated through NGOs has effectively quieted or neutralized popular protest before the public can reach state power.

Thus, through NGOs, hegemonic blocs are able to create the illusion that the distance between the poor and power is being bridged, when, in fact, it is increasing. The NGO phenomenon has thwarted the democratic process, because like corporations NGOs are ultimately accountable to their shareholders and sponsors, not the poor. Such an appropriation of resistance serves to complete the neo-liberal project begun by the World Bank and other actors. States also benefit from the presence of NGOs, especially those involved with aid and relief. Through NGOs, states have been able to renegotiate debts and lobby for aid, but such benefits seldom reach the poor. As noted in Chapter 1, the dependent articulation of Third World states to the global economy and constellations of power has necessitated the maintenance of conditions that would bring in foreign capital, even if those conditions spell a crisis for the poor.

If the World Bank were truly interested in the development of people, and not just markets, it would emphasize the catalytic role that states can play in development, rather than diminish those public functions. It is true, as Sen observes, that in many developing nations the state was involved in activities it did not always possess the competence for, such as running industries, while failing to do what it was expected to do, such as providing schools, health care, shelter, and land reform. The World Bank is correct in calling for the state's withdrawal from the former, according to Sen, but not in denying the state's responsibility to do the latter; in fact, the Bank has undermined the state's capacity to act on behalf of the poor in these areas.[1]

The erasure of national economic boundaries through free trade and the de-emphasis of the state is fatally wounding to "the major unit of community capable of carrying out any policies for the common good" (Daly 1994: 112). This includes not only national policies toward domestic ends, but also international agreements required for addressing those environmental problems that are irreducibly global (such as climate change and ozone depletion). Public institutions stabilize the market and help to prevent the criminal monopolization of critical distribution networks by establishing accountable alternatives. If it is truly interested in democracy, the World Bank ought to strengthen these public institutions wherever they exist, and establish them where they do not exist, instead of calling for their demise. Lester Pearson's (1969) well-known report, *Partners in Development* (discussed in Chapter 2), called for a partnership between multilateral development agencies and the developing world in addressing the socioeconomic needs of the latter. The World Bank, however, has rejected partnership and has chosen to become a "secular god" instead (Collier 1991: 111).

If the "dream" of a "poverty free world" is to be realized, the Bank ought to abandon its myopic obsession with the market. Peet appropriately recommends that "the poor should not be sentimentalized, but included in the process of planning their own development" (Peet 2003: 233).

Joseph Stiglitz (1998b: 42) called for a framework that "recognizes the importance of economic security and the creation of safety nets." He understood that "We need to move beyond measures of GDP and look at life spans and literacy rates" if any development is to occur (ibid.). However, as shown in the previous chapter, the attempts by Stiglitz and others to broaden the Bank's narrow, market-oriented focus have met with nothing but hostility and resistance. It seems that the "physicians" at the World Bank have dispensed the same neo-liberal nostrums to all of their "patients," regardless of their particular developmental ills, and plunged them into a downward spiral of debt, disease, poverty, and death. While the Bank has once again reiterated the need to address equity, poverty, and accountability, it remains to be seen whether the Bank's considerations of these issues will ever amount to anything more than a clever public relations ploy.

Apart from the World Bank, citizens and their governments ought to jointly develop methods of fighting poverty in their countries and ensure that their strategies are appropriate and sustainable if they are to resist World Bank interventions and appropriations. The national state ought to ensure that it represents its people, and that its policies serve their social, economic, political, cultural, environmental, and other developmental needs, if it wishes to retain its sovereignty against the assault of multilateral aid organizations.

Third World national states ought to question and resist blanket macroeconomic policies designed by foreign interests that fail to address the specific needs of their countries. Those external entities do not always have the best interest of the public in mind; they are obviously more interested in facilitating the growth of international capital, which has generated a serious conflict of interest. States and their civil societies must remember that golden straitjackets are still straitjackets. Developing countries may exercise greater agency if their states can achieve accountability with their civil societies. Together, they can decide what is the most appropriate relationship with external organizations for meeting their development needs. Without accountability, however, developing countries, especially the poor, will continue to be vulnerable to the manipulations of imperial and corporate interests. The growing influence of actors like the World Bank and the retreat of the state from its civic duties have only increased the distance between those who make decisions and those who must bear the consequences of those decisions.

Notes

I Theorizing the World Bank and development

1 The term "glocal" was coined by Roland Robertson in "Glocalization: Time–Space and Homogeneity–Heterogeneity," in Featherstone, M., Lash, S., and Robertson, R. (eds.) (1995), *Global Modernities*, London: Sage, pp. 25–44.
2 Personal correspondence with the author, November 11, 1991.

2 Toward social lending: shifts in the World Bank's development thinking

1 Opening remarks of Lord Keynes at the first meeting of the Second Commission on the Bank for Reconstruction and Development, US Department of State, *Proceedings and Documents of United Nations Monetary and Financial Conference*, Bretton Woods, NH, USA, July 1–22, 1945, vol. 1 (GPO 1948), Doc. 47, p. 84.
2 William Clark, a British national, was appointed Director of Information and Public Affairs at the World Bank by both presidents Woods and McNamara in April 1968. From 1974 until his retirement in 1980 he served as vice-president for external relations. Prior to that, Clark worked with Woods as director of the Overseas Development Institute (ODI) in London.
3 Third World delegates to Bretton Woods were mainly from Latin America. However, there was also an Indian delegation, although India was not an independent nation at the time of Bretton Woods. Most of Africa was still under direct colonial rule and not represented at the conference. The complete list of "governments and authorities" participating in the conference is: Australia, Belgium, Bolivia, Brazil, Canada, Chile, China, Colombia, Costa Rica, Cuba, Czechoslovakia, Dominican Republic, Ecuador, Egypt, El Salvador, Ethiopia, French Committee of National Liberation, Greece, Guatemala, Haiti, Honduras, Iceland, India, Iran, Iraq, Liberia, Luxembourg, Mexico, the Netherlands, New Zealand, Nicaragua, Norway, Panama, Paraguay, Peru, Philippine Commonwealth, Poland, Union of South Africa, Union of the Soviet Socialist Republics, United Kingdom, United States of America, Uruguay, Venezuela, and Yugoslavia.
4 See *Proceedings and Documents of United Nations Monetary and Financial Conference*, 1945, vol. 2, p. 1177, for a detailed discussion on the arguments of developing countries at the Bretton Wood Conference.
5 Reported in the *New York Times*, June 6, 1947.
6 Davidson Sommers came to the Bank from the US Department of Defense in 1946, served in the Bank's Legal Department, and then served as vice-president

until 1959. See Davidson Sommers, interview, Columbia University's World Bank Oral History Program, July 1985, pp. 1, 10–11.

7 See International Bank for Reconstruction and Development, *Fourth Annual Report 1948–1949*, Washington DC: World Bank, p. 47.

8 Eugene R. Black, "Address to the Board of Governors," in IBRD, Summary Proceedings, Eleventh Annual Meeting of the Board of Governors (November 15, 1956), p. 11.

9 Most of the Bank's loanable funds do not come from subscriptions at all. The Bank raises funds by selling its own bonds on the world financial markets and then charges its borrowers a marginally higher interest rate than it must pay its own bondholders. As the Bank's bonds are ultimately guaranteed by member governments, they are considered good investments and given a AAA grade by securities rating houses such as Moodys and Standard & Poor. Many institutional investors (like pension funds) buy them, as do individuals.

10 Robert W. Cavanaugh, interview, Columbia University's World Bank Oral History Program, July 25, 1961, pp. 63–4.

11 Robert Garner was Senior Vice-President at the Bank and a key assistant to President Eugene Black.

12 See World Bank (1953) *The Economic Development of Nicaragua*, Baltimore: Johns Hopkins University Press, pp. 22–3. (The mission chief was E. Harrison Clark.)

13 Eugene Black, address to the Investment Dealers' Association of Canada, Eastern District, Montreal, Canada, February 23, 1950. IBRD Press Release 173, pp. 12–13.

14 See IBRD, *Fifth Annual Report*, 1950, Washington DC: World Bank, p. 8.

15 Dulles served from 1953 to 1959 under President Dwight D. Eisenhower.

16 These platforms included the Sub-commission on Economic Development and the Economic and Employment Commission of the Economic and Social Council (ECOSOC), the Economic Committee of the Economic and Social Council, the Economic and Financial Committee of the General Assembly, and the Assembly itself.

17 United Nations Economic and Social Council, Report of the Third Session of the Sub-Commission.

18 Reported in footnote to paragraph 28 of the Sub-Commission's report. United Nations Economic and Social Council, *Report of the Third Session of the Sub-Commission on Economic Development*, Doc. E/CN.1/65 (April 12, 1949).

19 United Nations Secretariat, Department of Economic Affairs (1949) *Methods of Financing Economic Development in Under-Developed Countries*, New York: United Nations, p. 143.

20 United Nations Secretariat, Department of Economic Affairs (1949) *Methods of Financing Economic Development in Under-Developed Countries*, New York: United Nations.

21 Payer (1982) contends that the USA preferred the IDA to be under the auspices of the World Bank rather than the United Nations because the USA could assert greater control as the member with the highest number of votes at the World Bank. If the organization fell under the United Nations, it would be under the control of the General Assembly, where the USA had just one vote. Thus, once the need for an international development agency was agreed upon, the USA preferred such an agency to be part of the World Bank rather than the United Nations.

22 Annual Meeting of the Board of Governors, Washington DC, October 2, 1959. Remarks of President Eugene R. Black.

23 IBRD/IDA Press Release 67. Remarks at the Third Session of the Annual General Meeting by Isidro Ycaza Plaza, governor for Ecuador, October 1, 1959 (cited in Kapur *et al.* 1997).

24 C. Douglas Dillon was Undersecretary of State and alternative delegate of the Bank for the USA (cited in the Bank's Annual Report, September 30, 1959).

25 IBRD/IDA press release 67. Remarks at the Third Session of the Annual General Meeting by Isidro Ycaza Plaza, governor for Ecuador, October 1, 1959 (cited in Kapur *et al.* 1997).

26 Memorandum, The Bank's Policy Towards Municipal Water Supply Projects, October 5, 1960 (cited in Kapur *et al.* 1997).

27 Ibid.

28 Robert Garner, address to the board of governors, 1961 Annual Meeting, September 21, 1961, pp. 7–8.

29 Dillon also served as Kennedy's Secretary of the Treasury from 1961 to 1965.

30 Ward was the head of the International Institute for Environment and Development in London and lecturer at Harvard University. In the early Kennedy years, she and her husband, Robert Jackson, persuaded Kennedy to support Kwame Nkrumah, Ghana's president. She was also a close friend of Robert McNamara. The eclectic titles of her books reveal the range of her thought: *Development and Dependence in Emergent Africa* (1956); *India and the West* (1963); *The Rich Nations and Poor Nations* (1962); *Why Help India* (1963); *Women in the New Asia* (1965); *The Decade of Development* (1965); *Nationalism and Ideology* (1966); and *The Lopsided Earth* (1968).

31 Barbara Ward, address to the British Overseas Development Institute, June 6, 1965.

32 Higgins was a Canadian development economist who worked in Indonesia, taught at MIT, and wrote a widely used textbook on development economics.

33 McNamara writes in his book, *In Retrospect: The Tragedy and Lessons of Vietnam* (Time Books, 1995, p. 311), about his tenure as Secretary of Defense: "I do not know to this day whether I quit or was fired."

34 Personal interview with Kraske, September 22, 1994.

35 Speech to the American Society of Newspaper Editors, May 18, 1966. This speech was reprinted as the first essay in McNamara's book, *The Essence of Security* (McNamara 1968).

36 From testimony at a hearing of the Senate Foreign Relations Committee on June 14, 1969. *Congressional Quarterly Almanac*, 1969, p. 299.

37 McNamara's biographer, Shapley, found that McNamara had an obsession with numbers: "Numbers were a language for him. He frequently wanted numbers from his staff to the point that they were virtually forced to invent the desired figures rather than risk the president's wrath. At the Bank one still hears the story of McNamara calling a staffer back to Washington from his holiday in order to supply a single figure" (Shapley 1992: 608).

38 See McNamara's collection of speeches (McNamara 1981).

39 Mahbub ul Haq was an orthodox, growth-centered economist and planner in the early 1960s, who came to doubt this view by the late 1960s. He sent shockwaves through Karachi in 1968 by criticizing the concentration of wealth "in the hands of only twenty-two family groups," and the maldistribution of public services. See his work *The Poverty Curtain: Choices for the Third World* (ul Haq 1976), esp. pp. 3–11.

3 The search for an urban agenda at the World Bank

1 Interview with Michael Cohen, director of the World Bank's Urban Division, 1994. Michael Cohen began his career at the World Bank in 1972, when he started working in the Urban Division. He became head of the Urban Department in 1990 and served as senior advisor to the Bank's vice-president for environmentally sustainable development from 1994 to 1998.

2 A large part of the Bank's early lending was directed toward urban areas, especially power, water, and other infrastructure-related projects. However, until the 1970s, the Bank did not deal with the socioeconomic problems associated with rapid urbanization in Third World countries.

3 See Chapter 2 for a discussion on the IDA and the politics of its early lending.

4 Eugene R. Black, address to the Economic and Social Council of the United Nations, New York, April 24, 1961, p. 13.

5 Abrams was critical of the Bank's reasons for not funding housing programs. He noted that, while President Black's argument highlighted the magnitude of the problem and the financial limitations of multilateral agencies in funding housing, there were parallels with the argument in the USA in the 1930s against federal aid to housing when "it became fashion to use the adding machine as the field-piece against all social programs" (Abrams 1964: 96–7). According to Abrams, this was "neither good accounting nor good sense" because this approach confuses "loans with subsidies, then bundles them up together and multiplies them by thirty" (ibid.). Abrams believed that World Bank loans for housing ought to come out of a revolving fund composed of initial advances and repayments with interest, and that the total loans made ought to be only a fraction of the initial funds. Most of the capital, he continued, ought to be raised internally through the development of savings, though some external aid may be useful as a primer. For Abrams, the amount of external aid will depend on the type of housing, on whether the aid given will have multiplier effects, and whether that aid will develop local skills, materials and the use of local currencies.

6 See Chapter 2 for a discussion on trends in Bank lending.

7 See United Nations Mission of Experts on Tropical Housing (South and South-East Asia, November–December 1950 and January 1951). Report of the Secretary General (E/CN.5/251, March 1, 1951).

8 Report on International Housing Programs, Subcommittee on Housing, Committee on Banking and Currency, US Senate, 87th Congress, 2nd Session (Washington, 1962), p. 1.

9 Edward Jaycox began his career at the Bank in the mid-1960s. He was the first head of the newly established Urban Department in 1975. He served in various capacities at the World Bank, including vice-president for Africa, until his retirement in the mid-1990s.

10 David Henderson took over leadership of the Development Economics department in the late 1960s.

11 In order to structure the new initiatives of the Bank, McNamara created a "Special Projects Unit," which was headed by Robert Sadove, under the vice-president for economic research, Hollis Chenery.

12 Interview with Kenneth Bohr, 1995.

13 Kenneth Bohr (April 27, 1971), personal background notes.

14 Kenneth Bohr (May 14, 1971), "A Proposed Bank Approach," memorandum, p. 3.

15 Kenneth Bohr (May 13, 1971), "Key Problems," memorandum, p. 1.

16 Kenneth Bohr (May 13, 1971), "Key Problems," memorandum, p. 1.

17 Interview with Kenneth Bohr, 1995.

18 IBRD/IDA/IFC (International Finance Corporation), summaries of discussions at meeting of the executive directors, May 23, 1972, p. 8.

19 Keare was the first Director of the Urban and Regional Economics Division and principal author of the DED's *Housing: Sector Policy Paper* (World Bank 1975b). Keare's division eventually became a department with Keare as its head.

20 Henderson was the head of the DED.

21 Interview with Douglas Keare, 1995.

22　Keare was the principal author of this paper.
23　Interview with Douglas Keare, 1995.
24　Interview with Douglas Keare, 1995.
25　Interview with Douglas Keare, 1995.
26　Interview with Edward Jaycox, 1995. World Bank offices, Washington DC.
27　Memorandum from P.P.M. Cargill to Hollis Chenery, "Note on Urban Poverty,"
　　May 14, 1975.
28　Interview with Edward Jaycox, 1995.
29　Interview with Edward Jaycox, 1995.
30　Interview with Edward Jaycox, 1995.
31　Interview with Douglas Keare, 1995.
32　Interview with Edward Jaycox, 1995.
33　Otto Koenigsberger became prominent in development circles as the director
　　of housing for the Federal Government of India after independence. His office
　　was responsible for resettling and rehousing several million refugees in India
　　from Pakistan. He helped to establish the Government Housing Factory in Delhi
　　and planned six new towns for refugees, two of which were based on cooperative
　　ownership and self-help by the settlers. After his return to Europe in 1951,
　　Koenigsberger took up teaching at universities and began working closely with
　　various organizations of the United Nations. His chief concern over the next
　　thirty years was the need to establish a new body of knowledge for technical and
　　professional people in Third World countries. He is known for "action planning,"
　　which is his critical insight into the development process.
34　Interview with Edward Jaycox, 1995.
35　As noted earlier, the Bank did not publish world development reports in the
　　1970s. The presidential addresses at the annual meetings served as the principal
　　vehicle for the Bank's statements on new policy. The world development report
　　now sets the tone for the subject in discussion.
36　"Shelter projects" fell into two sub-categories. *Slum upgrading* was intended to
　　improve existing housing by providing residents with secure land tenure and
　　better access to credit for construction and by upgrading infrastructure such as
　　water supply, sewerage, electricity, roads, and sidewalks. *Sites-and-services* projects
　　were designed to encourage individuals to construct their own homes on serviced
　　sites, again by providing land tenure, access to loans, and a variety of essential
　　infrastructural and social services. "Urban transport projects" included efforts
　　to improve urban traffic management, upgrade bus and rail service in cities, and
　　improve roads and pedestrian walkways. "Integrated urban projects" were city-
　　wide investment programs or multisectoral projects that usually included more
　　ambitious transport and business support components than was possible in sub-
　　sectoral projects in shelter or transportation. "Regional development projects"
　　were efforts to extend the multisectoral approach of integrated urban projects
　　beyond individual cities to the region as a whole.
37　Interview with Kenneth Bohr, 1995.
38　Interview with Douglas Keare, 1995.
39　Interview with Michael Cohen, director of the World Bank's Urban Division,
　　1994.
40　Interview with Michael Cohen, director of the World Bank's Urban Division,
　　1994.
41　Project information cited from Bank appraisal reports.

4 The fall of poverty alleviation: the politics of urban lending at the World Bank

1 Interview with Michael Cohen, 1994.
2 Interview with Robert Buckley, principal economist, World Bank Urban Division, 1994.
3 Interview with Larry Hannah, economist, World Bank's Urban Division, 1994.
4 Interview with Larry Hannah, economist, World Bank's Urban Division, 1994.
5 Interview with Larry Hannah, economist, World Bank's Urban Division, 1994.
6 Ul Haq was one of McNamara's chief advisors.
7 Edward Jaycox later became vice-president for Africa.
8 Interview with Edward Jaycox, vice-president for Africa, 2001.
9 Interview with Edward Jaycox, vice-president for Africa, 2001.
10 Interview with Edward Jaycox, vice-president for Africa, 2001.
11 Interview with Edward Jaycox, vice-president for Africa, 2001.
12 Keare was the first Director of the Urban and Regional Economics Division and principal author of the DED's *Housing: Sector Policy Paper* (World Bank 1975b). Keare's division eventually became a department with Keare as its head.
13 Interview with Douglas Keare, 1994.
14 See World Bank (1992). This pamphlet describes the functions of the Urban Development Division.
15 Interview with Michael Cohen, 1994.
16 Interview with Douglas Keare, 1994.
17 In his memoirs, William Clark (1986: 263) notes that American support for McNamara's second term was late in coming, and that the USA supported him only after he had spoken personally with Nixon. The executive board of directors of the Bank also pressured the Nixon administration to support McNamara's second term. While the Carter administration was more supportive of the Bank, it had to deal with Congressional pressure for greater say in the operations of the Bank. See Schoultz (1982) for an analysis of this.
18 Peter T. Bauer was a conservative economist specializing in international markets and free economic development at the London School of Economics and Political Science, and a Fellow of the British Academy. He formerly taught at Cambridge University.
19 "Is the World Bank Biting Off More than it Can Chew?," *Forbes Magazine*, May 26, 1980.
20 "Blood and Treasure," *Barron's*, June 18, 1990.
21 Interview with Jochen Kraske, 1994.
22 Reported in the *New York Times*, "Foreign Aid: Debating Uses and Abuses," March 1, 1981, p. E5.
23 Stockman (1986) writes in his memoirs that Alexander Haig, the Secretary of State under Reagan, lobbied to continue US commitments to multilateral development agencies. Gwin (1994) notes that Jack Kemp, a staunch critic of foreign aid and the World Bank, negotiated a deal with Democrats that Republicans would support commitments to the IDA and World Bank if the Democrats supported Republican proposals for increased bilateral military assistance.
24 Opening remarks by President Ronald Reagan, Summary of Proceedings of the 1983 Annual Meeting of the Board of Governors, Washington DC.
25 Mahbub ul Haq, interview, World Bank Oral History Program, December 1982 (cited in Kapur *et al.* 1997).
26 Interview with Jochen Kraske, 1994.
27 Anne Krueger specialized in international economics and development. Her

research interests relate to policy reform in developing countries and US economic policy toward developing countries. She held the position of vice-president for economics and research at the Bank from 1982 to 1986. She is currently senior fellow at the Hoover Institution.

28 Reported in the *Sunday Times*, London, March 7, 1982, p. 10.
29 Reported in the *Sunday Times*, London, March 7, 1982, p. 10.
30 Reported in the *Sunday Times*, London, March 7, 1982, p. 10.
31 Ernest Stern made this recommendation in a review of the Bank's lending programs for the 1979 financial year. Stanley Please was a fiscal economist on the Bell mission to India in the 1960s. The mission's report (1968) recommended major macro-policy changes.
32 Telephone interview with Robert McNamara, 1993.
33 These figures were derived from World Bank annual reports from 1984 to 1994.
34 This report originated in 1979 when the African governors at the Bank asked the institution to undertake a special study of the region's needs and possibilities for accelerating its development. The report is frequently referred to as the Berg Report after is author, Elliot Berg, an economist from the University of Michigan.
35 Personal correspondence with the author. Barber B. Conable is frequently referred to as the "accidental president." When Clausen announced that he would not take up a second term, there was no obvious candidate to replace him. According to Kraske (1996), James Baker III and George Shultz were concerned that the impasse over Clausen's successor might lead the Europeans to name their own candidate. Baker nominated Conable purely as a tactical move and Conable himself had no desire to serve as president, having just retired after twenty years in Congress. However, Baker subsequently called Conable with the news that he was the only person they could agree on. After they appealed to his sense of public duty, Conable reluctantly accepted the position.
36 "The World Bank is Moving Away from Debt Policeman," *Weekly Mail and Guardian*, November 9, 1986.
37 Personal correspondence with Barber Conable, former president of the World Bank, 1995.
38 Speech at meeting to celebrate the Bretton Woods Conference, July 13, 1984, Bretton Woods, p. 4.
39 Ernest Stern's retirement speech at the Bank, January 26, 1995, (cited in Kapur *et al.* 1997).
40 Articles of Agreement (Article IV, Section 10).
41 Interview with Edward Jaycox, 2001.
42 Memo from Anthony Churchill to World Bank urban staff, April 30, 1981.
43 Interview with Anthony Churchill, 1995.
44 *Urban Strategy Paper*, 1981, World Bank (internal discussion document).
45 World Bank Office Memorandum from Joseph B. Buky to Hans Wyss, November 27, 1984.

5 Beyond global and local: a critical analysis of the World Bank and urban development in Zimbabwe

1 World Bank/IFC, Office Memorandum, Buky to Wyss, November 27, 1984.
2 Interview with Fred C. King, senior country officer, Southern Africa Department, World Bank, 1991.
3 Interview with Preben Jensen, principal highway engineer, Infrastructure and Operations Division of the Southern Africa Department, World Bank, 1991.

Jensen was involved in the early discussions on the Chitungwiza railway project.
4 Interview with Jeff Racki, 2001. Racki was a member of the first urban sector mission to Zimbabwe and later the task manager for the World Bank's first urban project in Zimbabwe.
5 In 1979, Britain brokered an agreement between the black nationalist movement and the white settlers that brought independence to Zimbabwe.
6 Mugabe's inaugural address, 1980 (www.polity.org.za/pol/speech/2003/?show=44745).
7 Fearing that Britain would grant independence to black Zimbabwe as it did to its other African colonies, white settlers called for a Unilateral Declaration of Independence for Rhodesia in 1965.
8 Interview with Diana Patel, Harare, Zimbabwe, 1993. Dr. Patel was then a lecturer in Sociology at the University of Zimbabwe. She has also written a book and a number of articles on low-income housing in Zimbabwe (see bibliography).
9 Ministry of Local Government, pamphlet, December 6, 1980.
10 Speech by Eddison Zvobgo, Minister of Local Government and Housing, January 21, 1981, Urban Councils Conference.
11 In the mid-1970s, Glen View was the first large area to be self-help built. Some 7,342 plots for homes with a wet core were provided by civil contractors. Applicants were first expected to have a monthly income of less than Z$100. The limit was subsequently raised to Z$150. The monthly payment for the plot was about Z$14. Loans for building materials were offered for up to Z$200 to each plot holder. Additional loans could be negotiated once the core of the house was constructed.
12 Reported in the *Herald*, Zimbabwe, January 11, 1983.
13 Reported in the *Herald*, Zimbabwe, September 24, 1981.
14 Reported in the *Herald*, Zimbabwe, June 19, 1982.
15 This settlement is located 10 km south-east of the Harare city center. The land used to belong to a Methodist missionary trust and was transferred to the state in the late 1970s. This settlement was upgraded, in part, because residents had legally paid rent and lived in the area for a long time.
16 Interview with Chris Mafico, deputy projects manager at Old Mutual Properties, Harare, Zimbabwe, 1993.
17 The Urban Sector Report (World Bank 1985b) was based on the findings of an urban sector mission that visited Zimbabwe from August 19 to September 5, 1981. The mission included P. Patel, J. Racki, G. Beier, J. Kozloswski, E. McKay, and E. Bachrach.
18 Interview with Jeff Racki, 2001.
19 Interview with Jeff Racki, 2001.
20 Interview with Jeff Racki, 2001.
21 Interview with Jeff Racki, 2001.
22 Interview with Diana Patel, Harare, Zimbabwe, 1993.
23 Interview with Middleton Nyoni, city treasurer, Bulawayo City Council, Zimbabwe, 2000.
24 Accusations of discrimination were reported in the following news reports: the *Herald*, Zimbabwe, July 14, 1981, April 13, 1982, and November 8, 1982; the *Sunday Mail*, Zimbabwe, December 9, 1982.
25 Interview with S. Chakaipa, 2000. Chakaipa was Deputy Secretary, Ministry of Local Government and Housing in Zimbabwe.
26 Reported in the *Herald*, Zimbabwe, May 6, 1980.
27 Reported in the *Herald*, Zimbabwe, March 4, 1980.
28 Interview with Jeff Racki, 2001.
29 Interview with James Hicks, 2001. Hicks was task manager for the World

Bank's second urban project, Urban II, and the regional development project in Zimbabwe.

30 Interview with James Hicks, 2001.
31 The Bank (World Bank 1989a: 57) specified the following criteria for towns that wished to participate in the project. (a) The council must demonstrate a satisfactory staffing capacity for managing the preparation and implementation of infrastructure projects. This should be in the form of a staffing plan, which could include reference to the technical assistance and training support provided under the project. The council's past record of performance would be taken into account. (b) The council must demonstrate satisfactory organizational and staffing arrangements for operating and maintaining its services. The staffing plan mentioned in (a) above should make reference to maintenance. Past performance with maintenance would be taken into account. (c) The council must have submitted to the Ministry of Local Government, Rural and Urban Development (MLGRUD) a financial recovery plan (FRP) that demonstrates that it will be able, during a period not exceeding five years, to eliminate all of its accumulated deficit, if any, and to bring each main account into surplus. The FPP must include an estimate of the financial impact of the proposed investment and it must indicate the source of funding of the additional recurrent costs that would be occasioned by the proposed investment. (d) The FPP must have been formally approved by the council, and endorsed by the MLGRUD, and a written undertaking provided to MLGRUD that the plan will be implemented without modification unless proper agreement is reached with the Ministry to change the plan. (e) The council must not be in arrears with its repayments to the General Development Loan Fund or to the NHF. (f) The council must demonstrate that within twelve months it will be current in the submission of its financial accounts for audit.
32 Interview with Jeff Racki, 2001.
33 Interview with James Hicks, 2001.
34 Interview with James Hicks, 2001.
35 Reported in the *Sunday Mail* (Zimbabwe), March 17, 1991.
36 Recorded in Hansard, Parliamentary Speech of B. Moyo, March 12, 1991.
37 Reported in the *Daily Mail and Guardian*, South Africa, July 19, 1999.
38 Interview with Colleen Butcher, Resident Representative at the World Bank Mission, Harare, 1993.
39 See http://allafrica.com/stories/200301040142.html for preliminary results of the Zimbabwe 2002 census.
40 Personal visit, 1993.
41 Reported in the *Sunday Mail*, Zimbabwe, August 14, 1995.
42 Reported in the *Herald*, Zimbabwe, September 20, 1994.
43 Personal interview, 1993.
44 Personal interview, 1993.
45 Interview with Colleen Butcher, 1993.
46 Personal interview, 1993.
47 Reported in the *Daily Mail and Guardian*, June 14, 2001.
48 *Parade*, Zimbabwe, April 1992.
49 "How Mujuru Made his Millions," *Horizon*, February 1993, pp. 6–7.
50 *Financial Gazette*, May 28, 1989.
51 "Major Disorders Cited In War Victims' Fund Claims," *Financial Gazette*, July 31, 1997. Also reported in an article by Farai Makotsi, "Ex-Combatants Cry Foul as Inquiry Unfolds," *Financial Gazette*, August 28, 1997.
52 See "Zim Ministers Looted War Veterans Fund," *Electronic Mail and Guardian*, April 22, 1997 (available at www.mg.co.za).

53 Interview with Colleen Butcher, 2001.
54 Reported in *Leadership Online,* December 8, 2000, vol. 18(1) (available at www. millennium.co.za).
55 "Mugabe's Unwanted Palace Taunts Food Rioters," *Sunday Times*, London, January 25, 1998.
56 "Mugabe's Unwanted Palace Taunts Food Rioters," *Sunday Times*, London, January 25, 1998.
57 Zimbabwe's (Government of Zimbabwe 1996) latest action plan for human settlements does not address the country's serious housing shortages. Instead, it rehashes failed policies, arriving at the pessimistic conclusion that the government's options are rather limited.
58 See www.architectafrica.com/bin1/zimbabwe080613.html
59 Reported by the BBC, June 20, 2005.
60 Reported in *Zwnews*, June 8, 2005.
61 Reported by the BBC, June 17, 2005.
62 Reported in the *Mail and Guardian*, June 19, 2005.
63 For an account of the World Bank's role in the policy choices of Zimbabwe, see Bond (1998).
64 World Bank Internal Office Memorandum from Joseph B. Buky to Hans Wyss, November 27, 1984.
65 Interview with James Hicks, 2001.
66 There are very few NGOs involved in the urban sector in Zimbabwe. Plan-International is the major one operating in Epworth, focusing on water reticulation work. Homeless International also does some work in this area.
67 Reported in the *Financial Gazette*, January 5, 1997.
68 Interview with Middleton Nyoni, 2000.

6 Globalization, neo-liberalism and the politics of the Bank's current urban agenda

1 See the Rainforest Action website (www.ran.org) for more information on its campaigns against the World Bank, and *New York Times*, May 1994 (various dates), as well as the advertisement.
2 Reported in the *Weekly Mail and Guardian*, November 1–5, 1994, p. 21
3 Reported in the *Weekly Mail and Guardian,* June 25 to July 1, 1999, p. 17.
4 Wolfensohn drafted the Comprehensive Development Framework in long hand at his vacation home in Jackson Hole, Wyoming. He circulated the draft of his framework in early 1999 and presented his ideas to the board of governors at the Bank's 1999 annual meeting.
5 Ravi Kanbur is T.H. Lee Professor of World Affairs and Professor of Economics at Cornell University. He was on the staff of the World Bank from 1989 to 1997, serving successively as economic adviser, senior economic adviser, resident representative in Ghana, chief economist of the African region of the World Bank, and principal adviser to the chief economist of the World Bank.
6 Kanbur consulted widely in drafting this report, traveling around the world to meet with NGOs and other civil society groups. He organized an internet conference in February 2000 to solicit comments on the draft report. Hundreds of individuals took part in the conference. The controversial, original draft of this report is available on various websites on the internet including Professor Kanbur's website (www.people.cornell.edu/pages/sk145/).
7 Lawrence Summers was a member of the MIT economics faculty. He was a domestic policy economist for President Reagan's Council of Economic

Advisors in 1981 and joined the Harvard economics faculty in 1982. In 1991 he was appointed as vice-president and chief economist for the World Bank. He became Under-Secretary of the Treasury in 1993, Deputy Secretary in 1995, and Secretary of the Treasury in July, 1999. He was appointed President of Harvard University in 2001 and resigned from the post in 2006.
8 David Moberg, "Silencing Joseph Stiglitz," *Salon News*, May 2, 2000.
9 David Moberg, "Silencing Joseph Stiglitz," *Salon News*, May 2, 2000.

7 Conclusion

1 Reported in the *Financial Times*, June 30, 2000.

Bibliography

Abrams, C. (1952) *Land Problems and Policies: Preliminary Analysis*, New York: United Nations.
—— (1964) *Man's Struggle for Shelter in an Urbanizing World*, Cambridge, MA: MIT Press.
Adegunleye, J.A. (1987) "The Supply of Housing, A Bid for Urban Planning and Development in Africa," *Habitat International* 11(2): 105–11.
Amin, S. (1972) "Underdevelopment and Dependence in Black Africa: Origins and Contemporary Forms," *Journal of Modern African Studies* 10(4): 503–24.
—— (1974) *Accumulation on a World Scale: A Critique of the Theory of Underdevelopment*, New York: Monthly Review Press.
—— (1990) *Maldevelopment: Anatomy of a Global Failure*, Atlantic Highlands, NJ: Zed Books.
Appadurai, A. (1990) "Disjuncture and Difference in the Global Cultural Economy," *Public Culture: Bulletin of the Project for Transnational Cultural Studies* 2(2): 4–6.
Astrow, A. (1983) *Zimbabwe: A Revolution that Lost its Way?*, London: Zed Books.
Ayres, R.L. (1983) *Banking on the Poor: the World Bank and World Poverty*, Cambridge: MIT Press.
Babai, D. (1984) *Between Hegemony and Poverty: The World Bank in the World Economy, 1944–1983*, unpublished doctoral thesis, University of California, Berkeley.
Baldwin, D.A. (1966) *Economic Development and American Foreign Policy 1943–1962*, Chicago: University of Chicago Press.
Bamberger, M., Sanyal, B., and Valverd, N. (1982) *Evaluation of Sites and Services Projects: The Experience from Lusaka, Zambia*, Washington DC: World Bank Staff Working Papers No. 548.
Bauer, P.T. (1979) "Commentary," in Ryan C. Amacher, Gottfried Haberler, and Thomas D. Willett (eds) *Challenges to a Liberal International Economic Order*, Washington DC: American Enterprise Institute, pp. 462–7.
Baum, W.C., and Tolbert, S.M. (1985) *Investing in Development: Lessons of the World Bank Experience*, New York: Oxford University Press (published for the World Bank).
Beauregard, R.A. (1984) "Structure, Agency and Urban Redevelopment," in Michael P. Smith (ed.) *Cities in Transformation*, Beverly Hills: Sage Publications, pp. 51–72.
—— (1995) "Theorizing the Global–Local Connection" in P.L. Knox and P.J. Taylor (eds) *World Cities in a World-System*, Cambridge: Cambridge University Press, pp. 232–48.

Beier, G., Churchill, A., Cohen, M., and Renaud, B. (1976) "The Task Ahead for the Cities of the Developing World," *World Development* 4(5): 363–409.

Beresford, W.M.P. (1992) "Low Income Home Loan Lending: The Zimbabwean Experience," paper presented to the Home Loan Guarantee Company "Strategic Think Tank," Johannesburg, July 9–10.

Berry, Brian J.L. (1970) *Geographical Perspectives on Urban Systems*, Englewood Cliffs, NJ: Prentice Hall.

—— (1981) *Comparative Urbanization: Divergent Paths in the Twentieth Century*, New York: St. Martin's Press.

Bhagwati, J.N. (1985) *Dependence and Interdependence*, Cambridge, MA: MIT Press.

Black, E.R. (1960) "Cyril Foster Lecture," *The Economic Journal* 70(278): 266–76.

Blitzer, S., Hardoy, J., and Satterthwaite, D. (1983) "The Sectoral and Spatial Distribution of Multilateral Aid for Human Settlements," *Habitat International* 7(1/2), 103–27.

Bloomstrom, M., and Hettne, B. (1984) *Development Theory in Transition*, London: Zed Press.

Bond, P. (1993) *Finance and Uneven Development in Zimbabwe*, unpublished PhD thesis, Johns Hopkins University, Baltimore.

—— (1998) *Uneven Zimbabwe: A Study of Finance, Development, and Underdevelopment*, Trenton: Africa World Press.

—— (2000) "Zimbabwe: Another Liberation?," *Indicator South Africa* 16(1): 95–103.

Brennan, E.M., and Richardson, H.W. (1986) "Urbanization and Urban Policy in Sub-Saharan Africa," *African Urban Quarterly* 1: 2–42.

Brickhill, J. (1999) "Mal de Mugabe," *Leadership Online Edition*, 17(2) (available at www.millennium.co.za).

Bruno, M. (1994) "Our Assistance Includes Ideas as Well as Money," *Transition, World Bank* 5(1): 1–4.

Bullard, N. (2000) "Another One Bites the Dust: Collateral Damage in the Battle for the Bank," Global Policy Forum (available at http://www.globalpolicy.org/socecon/bwi-wto/wbank/insidewb.htm).

Butcher, C. (1986) *Low Income Housing in Zimbabwe: A Case Study of the Epworth Squatter Upgrading Programme*, Occasional Paper Series, Harare: Department of Rural and Urban Planning.

Byres, T.J. (ed.) (1972) *Foreign Resources and Economic Development: A Symposium on the Report of the Pearson Commission*, London: Frank Cass.

Callaghy, T.M. (1986) *The Political Economy of African Debt: The Case of Zaire*, London: Macmillan.

Campbell, J. (1990) "World Bank Urban Shelter Projects in East Africa: Matching Needs With Appropriate Responses?," in P. Amis and P. Lloyd (eds) *Housing Africa's Urban Poor*, Manchester: Manchester University Press, pp. 205–26.

Carter Center (1990) *Beyond Autocracy in Africa*, Atlanta: The Carter Center of Emory University.

Castells, M. (1977) *The Urban Question*, Cambridge, MA: MIT Press.

—— (1998) *End of Millennium*, Oxford: Blackwell Publishers.

Castells, M., Goh, L., and Kwok, R.Y.W. (1990) *The Shek Kip Mei Syndrome*, London: Pion.

Caufield, C. (1996) *Masters of Illusion: The World Bank and the Poverty of Nations*, London: Pan Books.

Chase-Dunn, C. (1984) "Urbanization in the World-System: New Directions for Research," in M.P. Smith (ed.) *Cities in Transformation*, Beverley Hills, CA: Sage, pp. 111–20.

Chenery, H.B., and Strout, A.M. (1966) "Foreign Assistance and Economic Development," *American Economic Review* 56 (September): 679–733.

Chenery, H., Ahluwalia, M.S., Bell, C.L.G., Duloy J.H., and Jolly, R. (1974) *Redistribution with Growth: Policies to Improve Distribution in Developing Countries in the Context of Economic Growth: A Joint Study Commissioned by the World Bank's Development Research Center and the Institute of Development Studies, University of Sussex, London*, Oxford: Oxford University Press.

Clark, W. (1981) "Robert McNamara at the World Bank," *Foreign Affairs* 60 (1): 167–84.

—— (1986) *From Three Worlds (Memoirs)*, London: Sidgwick & Jackson.

Clausen, A.W. (1982a) "A Concluding Perspective," in Edward R. Fried and Henry D. Owen (eds) *The Future of the World Bank*, Washington DC: The Brookings Institution, pp. 67–81.

—— (1982b) "Global Interdependence in the 1980s," remarks before the Yomiuri Economic Society, Tokyo, Japan, January 13, 1982.

—— (1986) *The Development Challenge of the 1980s: Major Policy Address*, Washington DC: World Bank.

Clifton, D. (1992) (ed.) *Chronicle of the 20th Century*, Clifton, MO: J L International.

Cohen, J.M., Grindle, M.S., and Walker, T. (1985) "Foreign Aid and Conditions Precedent: Political and Bureaucratic Dimensions," *World Development* 13(12): 1211–30.

Cohen, M.A. (1983) *World Bank Lending for Urban Development, 1972–1982*, Washington DC: World Bank.

—— (1996) "The Hypothesis of Urban Convergence: Are Cities in the North and South Becoming More Alike in an Age of Globalization?," in Michael A. Cohen, Blair R. Ruble, Joseph S. Tulchin, and Allison M. Garland (eds) *Preparing for the Urban Future: Global Pressures and Local Forces*, Baltimore: Johns Hopkins University Press, pp. 25–38.

Collier, P. (1991) "From Critic to Secular God: the World Bank and Africa," *African Affairs* 90: 111–17.

Conteh-Morgan, E. (1990) *American Foreign Aid and Global Power Projection*, Brookfield, VT: Dartmouth.

Cox, R.W. (1993) "Gramsci, Hegemony, and International Relations: An Essay in Method," *Millennium: Journal of International Studies* 12(2): 162–210.

Cumming, S.D. (1990) "Post-Colonial Urban Residential Change in Zimbabwe: A Case Study," in R.B. Potter and A.T. Salau (eds) *Cities and Development in the Third World*, London: Mansell Publishing, pp. 32–50.

Currie, L. (1981) *The Role of Economic Advisors in Developing Countries*, Westport, CT: Greenwood Press.

Dalby, S. (1988) "Geopolitical Discourse: The Soviet Union as Other," *Alternatives* 13: 415–42.

—— (1990) *Creating the Second World War: The Discourse of Politics*, New York: Guilford Press.

Daly, H.E. (1994) "Farewell Lecture to the World Bank," in J. Cavanagh, D. Wysham, and M. Aruda (eds) *Beyond Bretton Woods: Alternatives to the Global Economic Order*, London: Pluto Press, pp. 109–17.

Dashwood, H.S. (1996) "The Relevance of Class to the Evolution of Zimbabwe's Development Strategy, 1980–1991," *Journal of Southern African Studies* 22(1): 27–48.

Davies, R.J. and Dewar, N. (1989) "Adaptive or Structural Transformation? The Case of the Harare, Zimbabwe, Housing System," *Social Dynamics* 15(1): 46–60.

Dear, M. (1986) "Post-Modern Planning," *Environment and Planning D: Space and Society* 4(3): 367–84.

Delorme, R. and André, C. (1983) *L'Etat et l'économie*, Paris: Senil.

Dengura, C. (1995) *The Case for Strategic Transportation Planning: A Focus on the Transportation Problem in Harare in High Density Suburbs*, Occasional Paper Series, Harare: Department of Rural and Urban Planning.

Department of Treasury (1982) *United States Participation in Multilateral Development Banks in the 1980s*, Washington DC: United States Department of Treasury, US Government Printing Office.

Domicelj, S. (1988) "International Assistance in the Urban Sector," *Habitat International* 12(2): 5–25.

Eisenhower, D.D. (1965) *Waging Peace: 1956–1961*, New York: Doubleday.

Eisenhower, M.S. (1965) *The Wine is Bitter: The United States and Latin America*, New York: Doubleday.

El-Shakhs, S. (1972) "Development, Primacy, and Systems of Cities," *Journal of Developing Areas* 7(1): 181–206.

Emmanuel, A. (1972) *Unequal Exchange*, London: New Left Books.

Escobar, A. (1995) *Encountering Development: The Making and Unmaking of the Third World*, Princeton, NJ: Princeton University Press.

Fainstein, S. (2001) "Inequality in Global City-Regions," in A.J. Scott (ed.) *Global City Regions: Trends, Theory, Policy*, London: Oxford University Press, pp. 285–98.

Fanon, F. (1963) *The Wretched of the Earth*, New York: Grove Press.

Fatton, Jr., R. (1989) "The State of African Studies and Studies of the African State: The Theoretical Softness of the Soft State," *Journal of Asian and African Studies* 24(3–4): 170–87.

Feinberg, R. (1988) "The Changing Relationship Between the World Bank and the International Monetary Fund," *International Organization* 42(3): 545–60.

Fichter, R., Turner, J.F.C., and Grenell, P. (1972) "The Meaning of Autonomy," in J.F.C. Turner and R. Fichter (eds) *Freedom to Build: Dweller Control of the Housing Process*, New York: Macmillan, pp. 122–47, 241–54.

Forbes, D., and Thrift, N. (eds) (1987) *The Socialist Third World: Urban Development and Territorial Planning*, New York: Basil Blackwell.

Foucault, M. (1978) *The History of Sexuality, I: An Introduction*, trans. R. Hurley, New York: Pantheon.

—— (1979) *Discipline and Punish*, New York: Vintage.

Frank, A.G. (1967) *Capitalism and Underdevelopment in Latin America*, London: Monthly Review Press.

—— (1969) *Latin America: Underdevelopment or Revolution*, New York: Monthly Review Press.

Friedman, T. L. (1999) *The Lexus and the Olive Tree*, New York: Farrar, Straus, and Giroux.

—— (2005) *The World is Flat: A Brief History of the Twenty-First Century*, New York: Farrar, Straus and Giroux.

Friedmann, J. (1966) *Regional Development Policy*, Cambridge, MA: MIT Press.

Fuchs, R.J. (1994) "Introduction," in R.J. Fuchs, E. Brennan, J. Chamie, and J.I. Uitto (eds) *Mega-City Growth and the Future*, Tokyo: United Nations University Press.

Fukuyama, F. (1992) *The End of History and the Last Man*, New York: The Free Press.

Gardner, R.N. (1969) *Sterling–Dollar Diplomacy: The Origins and Prospects of our International Economic Order*, New York: McGraw-Hill.

George, S., and Sabelli, F. (1994) *Faith and Credit: The World Bank's Secular Empire*, Boulder, CO: Westview Press.

Gibbon, P. (ed.) (1995) *Structural Adjustment and the Working Poor in Zimbabwe*, Uppsala: Reprocentralen HSC.

Gilbert, G., and Gugler, J. (1982) *Cities, Poverty, and Development*, Oxford: Oxford University Press.

Glassman, J. and Samatar, A. I. (1997) "Development Geography and the Third World State," *Progress in Human Geography* 21(2): 164–98.

Goodwin, R. (1988) *Remembering America*, Boston: Little Brown.

Gordon, D.F. (1984) "Development Strategy in Zimbabwe: Assessments and Prospects," in M. Schatzberg (ed.) *The Political Economy of Zimbabwe*, New York: Praeger, pp. 119–43.

Government of Zimbabwe (1981) *ZIMCORD Conference Documentation, March 23–27*, Salisbury, Zimbabwe.

—— (1982a) *Transitional National Development Plan, 1982/83–1984/85*, Harare: Government Printers.

—— (1982b) Remarks by Comrade S. Mumbengegwi, Meeting of the Executive Committee of the Local Government Association on Friday, October 15, 1982, Harare: Ministry of Housing.

—— (1985) *Long Term Plan 1985–2000*, Ministry of Public Construction and National Housing, Harare.

—— (1987) *Construction and Housing Cooperatives*, Harare, Zimbabwe: Ministry of Community and Cooperative Development.

—— (1998) *National Development Plan*, Harare: Government Printers.

—— (1991) *Zimbabwe: A Framework for Economic Reform, 1991–1995*. Harare: Government Printers.

—— (1992) *Census 1992: Zimbabwe Preliminary Report*, Harare: Government Printers.

—— (1996) *National Report and Plan of Action for Human Settlements in Zimbabwe*, Harare: Government Printers.

Gramsci, A. (1971) *Selections from the Prison Notebooks*, ed. and tr. Quentin Hoare and Godffrey Nowell Smith, New York: International Publishers.

Grimes, Jr., and Orville F. (1976) *Housing for Low Income Urban Families: Economics and Policy in the Developing World*, Baltimore: Johns Hopkins University Press.

Guarda, G.C. (1990) "A New Direction in World Bank Urban Lending to Latin American Countries," *Review of Urban and Regional Development Studies* 2(2): 115–25.

Gwin, C. (1994) *U.S. Relations with the World Bank*, Occasional Paper Series, Washington DC: The Brookings Institution.

Habermas, J. (1987a) *The Philosophical Discourse of Modernity*, Cambridge: Polity Press.

—— (1987b) *Theory of Communicative Action*, vols 1 and 2, Cambridge: Polity Press.

Hadwen, J.G., and Kaufmann, J. (1960) *How United Nations Decisions are Made*, London: A.W. Sijthoff.

Haggard, S. (1990) *Pathways from the Periphery*, Ithaca, NY: Cornell University Press.

Haggard, S. and Kaufman, R. (1992) "Introduction," in S. Haggard and R. Kaufman (eds) *The Politics of Structural Adjustment*, Princeton, NJ: Princeton University Press.

ul Haq, M. (1976) *The Poverty Curtain: Choices for the Third World*, New York: Columbia University Press.

—— (1998) *Reflections on Human Development*, Calcutta: Oxford University Press.

Harare (1985–95) *Annual Reports of the Department of Housing and Community Services*, Harare: City Council of Harare.

Harris, N. (1986) *The End of the Third World: Newly Industrializing Countries and the Decline of an Ideology*, London: I.B. Tauris.

—— (1989) "Aid and Urbanization: An Overview," *Cities*, August: 174–185.

Harris, R. (1997a) "The Silence of the Experts: Aided Self Help Housing, 1939–1954," unpublished paper, Department of Geography, McMaster University, Ontario Canada (available online www.science.mcmaster.ca/Geography/silence.html).

—— (1997b) "A Burp in Church, Jacob L. Crane's Vision of Aided Self-help Housing," *Planning History Studies* 11(1): 3–16.

Harvey, D. (1989) *The Condition of Postmodernity*, Cambridge: Basil Blackwell.

Havnevik, K.J. (ed.) (1987) *The IMF and the World Bank in Africa: Conditionality, Impact, and Alternatives*, Uppsala: Scandinavian Institute of African Studies.

Hayter, T. (1971) *Aid as Imperialism*, Harmondsworth: Penguin.

—— (1981) *The Creation of World Poverty*, London: Pluto.

Henderson, A.S. (2000) *Housing and the Democratic Ideal: The Life and Thought of Charles Abrams*, New York: Columbia University Press.

Herbst, J. (1989) "Political Impediments to Economic Rationality: Explaining Zimbabwe's Failure to Reform its Public Sector," *Journal of Modern African Studies* 27(1): 67–84.

Higgins, B. (1989) *The Road Less Travelled: A Development Economist's Quest*, Canberra: National Centre for Development Studies.

Hildebrand, M. (2001) "The Cities Alliance: A New Global Challenge to Urban Poverty," *Global Outlook*, January: 16–17.

Hoek-Smit, M. (1982) *Housing Preferences and Potential Housing Demand of Low-Income Urban Households in Zimbabwe*, Washington DC: United States Agency for International Development.

Holub, R. (1992) *Antonio Gramsci: Beyond Marxism and Postmodernism*, London: Routledge.

Horkheimer, M., and Adorno, T.W. (1974) *Dialectic of Enlightenment*, New York: Seabury.

Hyden, G. (1992) "Governance and the Study of Politics," in G. Hyden and M. Bratton (eds) *Governance and Politics in Africa*, Boulder, CO: Lynne Rienner Publishers.

Hyden, G., and Bratton, M. (eds) (1992) *Governance and Politics in Africa*, Boulder, CO: Lynne Rienner Publishers.

Iliffe, J. (1983) *The Emergence of African Capitalism*, London: Macmillan Press.

IBRD (International Bank for Reconstruction and Development) (1949) *Annual Report*, Washington DC: World Bank.

—— (1951) *Annual Report*, Washington DC: World Bank.

—— (1952) *The Economic Development of Jamaica: Report by a Mission of the International Bank for Reconstruction and Development*, Baltimore: Johns Hopkins University Press.

—— (1961) *Annual Report*, Washington DC: World Bank.

—— (1981) *Annual Report*, Washington DC: World Bank.

ILO (International Labor Organization) (1976) *Employment, Growth and Basic Needs: A One World Problem*, Geneva: ILO.

Jackson, R.H. (1977) *Plural State and New Societies*, Berkeley: University of California.

Jaycox, E. (1978) "The World Bank and Urban Poverty," *Finance and Development* 15(3): 10–13.

Jere, H. (1984) "Lusaka: Local Participation in Planning and Decision Making," in G.K. Payne (ed.) *Low Income Housing in the Developing World*, New York: John Wiley and Sons, pp. 55–68.

Jones, G.A., and Ward P.M. (1995) "The Blind Men and the Elephant: A Critic's Reply," *Habitat International* 19(1): 61–72.

Kahler, M. (1992) "External Influence, Conditionality, and the Politics of Adjustment," in S. Haggard and R. Kaufman (eds) *The Politics of Structural Adjustment*, Princeton, NJ: Princeton University Press, pp. 89–136.

Kamete, A.Y. (1999) "Government, Building Societies and Civil Society: Exploring the Interface in the Context of Low-Income Housing in Zimbabwe," *African Review of Money Finance and Banking, Supplementary Issue on Savings and Development*, pp. 77–96.

—— (2001) "The Quest for Affordable Urban Housing: A Study of Approaches and Results in Harare, Zimbabwe," *Development Southern Africa* 18(1): 31–44.

Kanbur, R. (1999) "The Strange Case of The Washington Consensus: A Brief Note on John Williamson's 'What Should the Bank Think about the Washington Consensus?'" An amended version of comments made at a PREM Week panel in the World Bank in July 1999 (available at http://people.cornell.edu/pages/sk145/papers/Washington%20Consensus.pdf).

—— (2000) *World Development Report*, draft version.

Kapur, D., Lewis, J.P., and Webb, R. (1997) *The World Bank: Its First Half Century*, Washington DC: The Brookings Institution Press.

Keare, D. (1983) "Affordable Shelter and Urban Development: 1972–82," *World Bank Research News*, Summer: 3–14.

Keare, D., and Parris, S. (1982) *Evaluation of Shelter Programs for the Urban Poor*, World Bank Staff Working Papers No. 547, Washington DC: World Bank.

Keyes, W.J., and Burcoff, M.C.R. (1976) *Housing the Urban Poor: Non-Conventional Approaches to a National Problem*, Institute of Philippine Culture, Poverty Research Series, No. 4, Quezon City.

Keyman, E.F. (1997) *Globalization, State, Identity/Difference: Toward a Critical Social Theory of International Relations*, Atlantic Highlands, NJ: Humanities Press.

Kincaid, A.D., and Portes, A. (1994) *Comparative National Development*, Chapel Hill, NC: University of North Carolina Press.

Knight, R.V. (1989) "The Emergent Global Society," in R.V. Knight and G. Gappert (eds) *Cities in a Global Society*, Newbury Park, CA: Sage Publications, pp. 24–43.

Koenigsberger, O.H., Groak, S., and Berstein, B. (1980) *The Work of Charles Abrams: Housing and Urban Renewal in the USA and the Third World*, New York: Pergamon Press.

Kolko, G. (1988) *Confronting the Third World: United States Foreign Policy, 1945–1980*, New York: Pantheon.

Kraske, J. (1996) *Bankers with a Mission: Presidents of the World Bank 1946–1991*, New York: Oxford University Press (published for the World Bank)

Laclau, E. and Mouffe, C. (1985) *Hegemony and Socialist Strategy: Towards a Radical Democratic Politics*, London: Verso.

LaFeber, W. (1993) *America, Russia, and the Cold War 1945–1992*, New York: McGraw Hill.

Lancaster, C. (1987) "The World Bank in Africa Since 1980: The Politics of Structural Adjustment Lending," in D. Kapur, J.P. Lewis, and R. Webb (eds) *The World Bank: Its First Half Century*, Washington DC: The Brookings Institution, pp. 161–94.

Laquian, A.A. (1977) "Whither Site and Services?," *Habitat International* 2(3/4): 291–301.

—— (1983a) *Basic Housing: Policies for Urban Sites, Services, and Shelter in Developing Countries*, Ottawa, Canada: International Development Research Center.

—— (1983b) "Sites, Services and Shelter – an Evaluation," *Habitat International* 5(5/6): 211–25.

Lefebvre, H. (1991) *The Production of Space*, Oxford: Basil Blackwell.

Linn, J.F. (1983) *Cities in the Developing World*, New York: Oxford University Press.

Lipietz, A. (1986) "New Tendencies in the International Division of Labor: Regimes of Accumulation and Modes of Regulation," in A.J. Scott and M. Storper (eds) *Production, Work, Territory*, Boston: Allen & Unwin.

Lipietz, A. (1987) *Mirages and Miracles*, New York: Verso.

Lipton, M. (1976) *Why Poor People Stay Poor: Urban Bias in World Development*, Cambridge: Harvard University Press.

Lipton, M., and Paarlberg, R.L. (1990) *The Role of the World Bank in Agricultural Development in the 1990s*, Washington, DC: International Food Policy Research Institute.

Ljung, K. (1990) "The Role of the World Bank in Housing," *Interplan* (American Planning Association) 33 (Fall): 1–4.

Mabogunje, A.L. (1981) *The Development Process: A Spatial Perspective*, New York: Holmes & Meier Publishers.

McCarney, P.L. (1987) *The Rise and Fall of Sites and Services at the World Bank*, unpublished PhD thesis, MIT, Cambridge, MA.

McCormick, T J. (1989) *America's Half Century: United States Foreign Policy and the Cold War*, Baltimore: Johns Hopkins University Press.

McGee, T.G. (1971) *The Urbanization Process in the Third World*, London: Bell and Sons.

McNamara, R.S. (1968) *The Essence of Security: Reflections in Office*, New York: Harper and Row.

—— (1981) *The McNamara Years at the World Bank: The Major Policy Addresses of Robert S. McNamara*, Baltimore: Johns Hopkins University Press.

McNeill, D. (1981) *The Contradictions of Foreign Aid*, London: Croom Helm.

Maddux, J.L. (1981) *The Development Philosophy of Robert S. McNamara*, Washington DC: World Bank.

Mafico, C.J.C. (1985) *An Evaluation of Urban Planning Standards for Low Income Housing in Zimbabwe*, Occasional Paper Series, Harare: Department of Rural and Urban Planning.

—— (1991) *Urban Low-Income Housing in Zimbabwe*, Brookfield, VT: Gower Publishing.

Mallaby, S. (2004) *The World's Banker: A Story of Failed States, Financial Crises, and the Wealth and Poverty of Nations*, New York: Penguin Press.

Mandaza, I. (1986) *Zimbabwe: The Political Economy of Transition*, Dakar, Senegal: CODESRIA.

Mandel, E. (1980) *Long Waves of Capitalist Development: The Marxist Interpretation*, Cambridge: Cambridge University Press.

Marx, K., and Engels, F. (1989) *The Communist Manifesto*, New York: International Publishers.

Mason, E.S., and Asher, R.E. (1973) *The World Bank Since Bretton Woods*, Washington DC: The Brookings Institution.

Massey, D. (1992) "Politics and Space/Time," *New Left Review* 196: 65–84.

—— (1993) "Power-Geometry and a Progressive Sense of Place," in J. Bird, B. Curtis, T. Putnam, G. Robertson, and L. Tickner (eds) *Mapping the Futures – Local Cultures, Global Change*, London: Routledge.

Mayo, S.K., Malpezzi, S., and Gross, D.J. (1986) "Shelter Strategies for the Urban Poor in Developing Countries," *World Bank Research Observer* 1(2): 183–204.

Merrifield, A. (2002) *Dialectical Urbanism: Social Struggles in the Capitalist City*, New York: Monthly Review Press.

Mghweno, J. (1984) "Tanzania's Surveyed Plots Programme," in G.F. Payne (ed.) *Low-income Housing in the Developing World*, New York: John Wiley and Sons, pp. 109–23.

Milder, D.C. (1996) "Foreign Assistance: Catalyst for Domestic Coalition Building," in J.M. Griesgraber and B.G. Gunter (eds) *The World Bank: Lending on a Global Scale*. London: Pluto Press; Washington DC: Center for Concern, pp. 142–91.

Milkman, R. (1979) "Contradictions of semi-peripheral development: the South African case," in W.L. Goldfrank (ed.) *The World System of Capitalism: Past and Present*, Beverly Hills, CA: Sage Press.

Miller-Adams, M. (1999) *The World Bank: New Agendas in a Changing World*, London: Routledge.

Moll, T. (1988) "The Limits of the Possible: Macro-Economic Policy and Income Redistribution in Latin America and South Africa," in J. Suckling and L. White (eds) *After Apartheid: Renewal of the South African Economy*, Trenton: Africa World Press.

Möller, V. (1974) "Some Aspects of Mobility Patterns of Urban Africans in Salisbury," *Proceedings of the Geographical Association of Rhodesia* 7: 22–32.

Morss, E. (1991) "The New Global Players: How They Compete and Collaborate," in *World Development* 19(1), 55–64.

Mosley, P., and Eeckhout, M.J. (2000) "From Project Aid to Programme Assistance," in Finn Tarp (ed.) *Foreign Aid and Development*, New York: Routledge, pp. 131–53.

Mosley, P., Harrigan, J., and Toye, J. (1991) *Aid and Power: The World Bank and Policy Based Lending*, New York: Routledge.

Mugabe, R. (1986) *Opening Address, Proceedings of the 10th Conference on Housing and Urban Development in Sub-Saharan Africa*, Harare: United States Agency for International Development.

Musanda-Nyamayaro, O. (1993) "Housing Design Standards for Low-Income People in Zimbabwe," *Third World Planning Review* 15(4): 329–54.

Musiyazviriyo, R. (1992) *An Analysis of the Bus Public Transportation Service of the City of Harare*, MA thesis, University of Zimbabwe, Harare.

Mutizwa-Mangiza, N.D. (1986) "Urban Centres in Zimbabwe: Inter-censal Changes, 1962–1982," *Geography* 71(2): 148–51, 311.

—— (1991) "Financing Urban Shelter Development in Zimbabwe: A Review of Existing Institutions, Problems and Prospects," *Habitat International* 15(1/2): 51–68.

Myrdal, G. (1968) *Asian Drama: An Inquiry into the Poverty of Nations*, Harmondsworth, Penguin.

Nelson, P.J. (1995) *The World Bank and Non-Governmental Organizations*, New York: St. Martin's Press.

Nyangoni, C., and Nyandoro, G. (1979) *Zimbabwe Independence Movements: Selected Documents*, London: Rex Collings.

O'Brien, R. (1992) *Global Financial Integration: The End of Geography*, London: Pinter.

O'Connor, A. (1986) *The African City*, London: Hutchinson.

Ohmae, K. (1990) *The Borderless World: Power and Strategy in the Interlinked Economy*, London: HarperCollins.

Onibokun, A.G., Agbola, T., and Labeodan, O. (1989) "World Bank Assisted Sites-and-Services Projects: Evaluation of Nigeria's Experiment," *Habitat International* 13(3): 51–64.

Packenham, R. (1973) *Liberal America and the Third World*, Princeton, NJ: Princeton University Press.

Palmer Associates (1995) *Report on Zimbabwe's Private Sector Housing Program*, Harare: Palmer Associates.

Parsons, T. (1951) *The Social System*, Glencoe, IL: Free Press.

Patel, D. (1984) "Housing the Urban Poor in the Socialist Transformation of Zimbabwe," in M. Schatzberg (ed.) *The Political Economy of Zimbabwe*, New York: Praeger, pp. 182–96.

—— (1988) "Some Issues of Urbanization and Development in Zimbabwe," *Journal of Social Development in Africa* 3(2): 17–31.

Patel, D., and Adams, R.J. (1981) *Chirambahuyo: A Case Study in Low-Income Housing*, Gweru: Mambo Press.

Paul, S. (1991) "Non-Governmental Organizations and the World Bank: An Overview," in Samuel Paul and Arturo Israel (eds) *Non-Governmental Organizations and the World Bank: Cooperation for Development*, Washington DC: World Bank, pp. 1–19.

Payer, C. (1974) *The Debt Trap: The I.M.F. and the Third World*, New York: Monthly Review Press.

—— (1982) *The World Bank: A Critical Analysis*, New York: Monthly Review Press.

Pearson, L B. (1969) *Partners in Development: Report of the Commission on International Development*, New York: Praeger Publications.

Peet, R. (2003) *Unholy Trinity: The IMF, World Bank, and WTO*, London: Zed Books.

Perlman, J. (1971) *The Fate of Migrants in Rio's Favelas: The Myth of Marginality*, PhD dissertation, MIT, Cambridge, MA.

—— (1976) *The Myth of Marginality: Urban Poverty and Politics in Rio de Janeiro*, Berkeley: University of California Press.

—— (1980) "The Failure of Influence: Squatter Eradication in Brazil," in M.S. Grindle (ed.) *Politics and Policy Implementation in the Third World*, Princeton, NJ: Princeton University Press, pp. 250–78.

Phaup, D.E. (1984) *The World Bank: How it Can Serve U.S. Interests*, Washington DC: The Heritage Foundation.

Pickard-Cambridge, C. (1988) *Sharing the Cities: Residential Desegregation in Harare, Windhoek and Mafikeng*, Johannesburg: South African Institute of Race Relations.

Pithouse, R. (2003) "Producing the Poor: The World Bank's New Discourse of Domination," *African Sociological Review* 7(2): 1–26.

Please, S. (1984) *The Hobbled Giant: Essays on the World Bank*, Boulder, CO: Westview Press.

Potter, R.B. (1985) *Urbanization and Planning in the Third World*, London: Croom Helm.

Potts, D., with Mutambirwa, C.C. (1991) "High Density Housing in Harare: Commodification and Overcrowding," *Third World Planning Review* 13(1): 1–25.

Poulantzas, N. (1978) *State, Power, Socialism*, London: Verso.

Prazniak, R., and Dirlik, A. (2001) *Places and Politics in an Age of Globalization*, New York: Rowman and Littlefield.

Pugh, C. (1988) "World Bank Housing Policy in Madras," *Habitat International* 12(4): 29–44.

—— (1989a) "Housing Policy Reform in Madras and the World Bank," *Third World Planning Review* 11(3): 247–73.

—— (1989b) "The World Bank and Urban Shelter in Bombay," *Habitat International* 13(3): 23–49.

—— (1994) "Housing Policies in Developing Countries: The World Bank and Internationalization, 1972–1993," *Cities* 11(3): 159–80.

Raffer, K. (1999) "More Conditions and Less Money: Shift of Aid Policies During the 1990s," paper presented at the Annual Meeting of the Developmental Studies Association, University of Bath, September 12–14.

Rakodi, C. (1987) *Housing and Urban Development in Lusaka, An Evaluation of Squatter Upgrading in Chawana*, unpublished PhD thesis, University of Wales.

—— (1989) "Housing Production in Harare Zimbabwe: Components, Constraints and Policy Outcomes," *Trialog* 20: 14–30.

—— (1995) "Housing Finance for Lower Income Urban Households in Zimbabwe," *Housing Studies* 10(2): 199–227.

—— (1997) "Residential Property Markets in African Cities," in Carole Rakodi (ed.) *The Urban Challenge in Africa*, New York: United Nations University Press, pp. 371–410.

Rakodi, C., and Mutizwa-Mangiza, N.D. (1989) *Housing Policy, Production and Consumption: A Case Study of Harare*, RUP Teaching Paper No. 3, Harare: Department of Rural and Urban Planning.

Rakodi, C. and Withers, P. (1995) "Home Ownership and Commodification of Housing in Zimbabwe," *International Journal of Urban and Regional Research* 19(2): 250–71.

Ravenhill, J. (ed.) (1986) *Africa in Economic Crisis*, London: Macmillan.

Reed, W.C. (1987) "Global Incorporation, Ideology, and Public Policy in Zimbabwe," *African Studies Review* 15: 49–59.

Renaud, B.M. (1983) *National Urbanization Policies in Developing Countries*, New York: Oxford University Press.

Richardson, H.W. (1987a) "Whither National Urban Policy in Developing Countries?," *Urban Studies* 24: 227–44.

—— (1987b) "Spatial Strategies, the Settlement Pattern, and Shelter and Service Policies," in Lloyd Rodwin (ed.) *Shelter, Settlement and Development*, Boston: Allen & Unwin, pp. 207–35.

Rimmer, D. (1981) "Basic Needs and the Origins of the Development Ethos," *Journal of Developing Areas* 15 (January): 220–32.

Robertson, R. (1995) "Glocalization: Time-Space and Homogeneity–Heterogeneity" in M. Featherstone, S. Lash, and R. Robertson, (eds) *Global Modernities*, London: Sage Publications, pp. 25–44.

Rodney, W. (1972) *How Europe Under-Developed Africa*, Washington DC: Howard University Press.

Rosenstein-Roden, P. (1957) *The Objectives of United States Economic Assistance Programs*, Cambridge, MA, Massachusetts Institute of Technology, Center for International Studies.

Ross, M.H. (1973) "Community Formation in an Urban Squatter Settlement," *Comparative Political Studies* 6: 296–316.

Rostow, W.W. (1960) *The United States in the World Arena; an Essay in Recent History*, New York: Harper & Row.

—— (1964) *The Process of Economic Growth*, New York: Norton.

Rowen, H. (1994) *Self-Inflicted Wounds*, New York: Random House.

Said, E. (1983) *The World, the Text, and the Critic*, Cambridge, MA: Harvard University Press.

Sandström, S. (2000) Opening Address, Global Conference on Capital Markets Development at the Sub-national Level, New York, February 15–17, 2000.

Sanyal, B. (1986) "Learning Before Doing: A Critical Evaluation of the Privatization Concept in Shelter Policies of International Institutional Institutions," *Open House International* 11(4): 13–21.

SAPRIN (Structural Adjustment Participatory Review International Network) (2004) *Structural Adjustment: The SAPRI Report*, London: Zed Books.

Schlyter, A. (1985) "Housing Strategies: The Case of Zimbabwe," *Trialog* 6: 20–9.

—— (1990) "Zimbabwe," in K. Mathey (ed.) *Housing Policies in the Socialist Third World*, London: Mansell Publishing, pp. 197–225.

Schmidt, W.E. (1979) "Rethinking the Multilateral Development Banks," *Policy Review* 10 (Fall): 47–61.

Schoultz, L. (1982) "Politics, Economics and U.S. Participation in Multilateral Development Banks," *International Organization* 36(3): 537–74.

Seers, D. (1969) "The Meaning of Development," *International Development Review* 11: 2–6.

Sen, A. (1999) *Development as Freedom*, New York: Knopf Publishers.

Shafer, D.M. (1986) "Undermined: The Implications of Mineral Export Dependence for State Formation in Africa," *Third World Quarterly* 8(3): 916–52.

Shafer, D.M. (1994) *Winners and Losers*, Ithaca, NY: Cornell University Press.

Shapley, D. (1992) *Promise and Power: The Life and Times of Robert McNamara*, Boston: Little Brown and Co.

Shihata, I. (1991) *The World Bank in a Changing World: Selected Essays*, Boston: Martinus Nijhoff Publishers (published for the World Bank).

Shultz, T. (1961) "Investment in Human Capital," *American Economic Review* 51: 3–24.

SIDA (Swedish International Development Agency) (1995) *Towards an Urban World: Urbanization and Development Assistance*, Stockholm: SIDA.

Sigmund, P.E. (ed.) (1972) *The Ideologies of the Developing Nations*, New York: Praeger Publications.

Silas, J. (1984) "The Kampung Improvement Project of Indonesia," in Geoffrey K. Payne (ed.) *Low Income Housing in the Developing World*, New York: John Wiley and Sons, pp. 69–88.

Simon, D. (1992) *Cities, Capital and Development: African Cities in the World Economy*, London: Belhaven Press.

Singer, H. (1964) "Education and Economic Development," in Hans Singer (ed.) *International Development: Growth and Change*, New York: McGraw Hill, pp. 36–49.

—— (1976) "Early Years, 1910–1983," in Alec Cairncross and Mohinder Puri (eds) *Employment, Income Distribution and Development Strategy: Problems of Developing Countries, Essays in Honor of H.W. Singer*, Holmes and Meier, pp. 70–82.

Sivaramakrishnan, A. (2000) "The Future of Development Assistance," *UN Chronicle* 37(4): 23–35.

Skocpol, T. (1977) "Wallerstein's World Capitalist System: A Theoretical and Historical Critique," *American Journal of Sociology* 82: 1075–90.

—— (1979) *States and Social Revolutions*, New York: Cambridge University Press.

—— (1987) "The Dead End of Meta-Theory," *Contemporary Sociology* 16: 1078–86.

Smith, N. (1984) *Uneven Development*, Oxford: Basil Blackwell.

—— (1993) "Homeless/Global: Scaling Places," in J. Bird, B. Curtis, T. Putnam, G. Robertson, and L. Tickner (eds) *Mapping the Futures – Local Cultures, Global Change*, London: Verso Press, pp. 219–45.

—— (2002) "New Globalism, New Urbanism: Gentrification as Global Urban Strategy," *Antipode* 3: 427–49.

Sorensen, T. (ed.) (1988) *Let the Word Go Forth: The Speeches, Statements, and Writings of John F. Kennedy*, New York: Delacorte Press.

Srinivasan, T.N. (1977) "Development, Poverty, and Basic Human Needs: Some Issues," *Food Research Institute Studies,* 16(2): 12–23.

Stallings, B. (1992) "International Influence on Economic Policy: Debt, Stabilization, and Structural Reform," in S. Haggard and R. Kaufman (eds) *The Politics of Structural Adjustment*, Princeton, NJ: Princeton University Press, pp. 41–88.

Steinberg, S. (1995) *Turning Back: The Retreat From Racial Justice in American Thought and Policy*, Boston: Beacon Press.

Stiglitz, J. (1998a) "More Instruments and Broader Goals: Moving Towards the Post Washington Consensus," Annual WIDER lecture, Helsinki, Finland.

—— (1998b) "Towards a New Paradigm for Development: Strategies, Policies and Processes," The Paul Prebisch Lecture, UNCTAD, Geneva.

—— (2000a) "The Paternalistic Attitude of the North Must Change: Why Joseph Stiglitz Retired from the World Bank," *Development and Cooperation* 2: 26–7.

—— (2000b) "What I Learned at the World Economic Crisis: The Insider," *The New Republic*, April 17/24: 56–60.

—— (2002) *Globalization and its Discontents*, New York: Norton.

Stockman, D.A. (1986) *The Triumph of Politics: How the Reagan Revolution Failed*, New York, Harper & Row.

Stoneman, C. (1993) "The World Bank: Some Lessons for South Africa," *Review of African Political Economy* 58: 87–98.

—— (1999) "Zimbabwe: A Good Example Defused," *Indicator SA* 15(7): 77–81.

Stoneman, C., and Cliffe, L. (1989) *Zimbabwe: Politics, Economics and Society*, London: Pinter Publishers.

Strange, S. (1996) *The Retreat of the State: The Diffusion of Power in the World Economy*, Cambridge: Cambridge University Press.

Streeten, P. (1977) *The Distinctive Features of a Basic Needs Approach to Development*, Washington DC: World Bank Policy Planning and Program Review Department.

Streeten, P., Burki, S.J., ul Haq, M., Hicks, N. and Stewart, F. (1981) *First Things First: Meeting Basic Human Needs in the Developing Countries*, New York: Oxford University Press.

Stren, R.E. (1978) *Housing the Urban Poor in Africa: Policy, Politics, and Bureaucracy in Mombasa*, Berkeley: Institute of International Studies.

—— (1994) "Towards a Research Agenda for the 1990s: An Introduction," in Richard Stren (ed.) *Urban Research in the Developing World: Africa*, Toronto: University of Toronto Press.

Swyngedouw, E.A. (1992a) "Elements of a Regulation Approach of Regional Development and Restructuring," paper presented at the Annual Conference of the American Association of Geographers, Baltimore.

—— (1992b) "The Mammon Quest: 'Glocalization', Interspatial Competition and the Monetary Order: The Construction of New Scale," in M. Dunford and G. Kafkalas (eds) *Cities and Regions in the New Europe*, London: Belhaven Press, pp. 36–97.

—— (1997) "Neither Global nor Local: Glocalization and the Politics of Scale," in K.R. Cox (ed.) *Spaces of Globalization: Reasserting the Power of the Local*, New York: Guilford Press, pp. 137–66.

Sylvester, C. (1991) *Zimbabwe: The Terrain of Contradictory Development*, Boulder, CO: Westview Press.

Taper, B. (1980) "Charles Abrams in Ghana," *Habitat International* 5(1/2): 49–53.

Taylor, P.J. (1985) *Political Geography: World Economy, Nation-State and Locality*, Harlow: Longman.

Teedon, P.L. (1990) *An Analysis of Aided Self-Help Housing Schemes: A Study of a Former Colonial City, Harare, Zimbabwe*, unpublished doctoral thesis, University of Keele.

Tevera, D. (1995) "The Medicine that Might Kill the Patient: Structural Adjustment and Urban Poverty in Zimbabwe," in David Simon, Wim Van Spengen, Chris Dixon, and Anders Närman (eds) *Structurally Adjusted Africa: Poverty, Debt and Basic Needs*, London: Pluto Press, pp. 79–90.

Thompson, J.B. (1984) *Studies in the Theory of Ideology*, Berkeley: University of California Press.

Todaro, M.P. (1977) *Economic Development in the Third World*, New York: Longman.

Torrie, J. (ed.) (1983) *Banking on Poverty: The Global Impact of the IMF and the World Bank*, Toronto: Between the Lines Press.

Tuck-Primdahl, M. (1991) "The World Bank Shifts Urban Policy," *The Urban Edge: Issues and Innovations* 15(2): 1–7.

Turner, J.F.C. (1963) "Village Artisan's Self-Built House," *Architectural Design* 33: 361–2.

—— (1965) "Lima's Barriadas and Corralones: Suburbs versus Slums," *Ekistics* 19: 152–5.

—— (1972a) "The Re-Education of a Professional," in J.F.C. Turner and R. Fichter (eds) *Freedom to Build: Dweller Control of the Housing Process*, New York: Macmillan, pp. 122–47.

—— (1972b) "Housing as a Verb," in J.F.C. Turner and R. Fichter (eds) *Freedom to Build: Dweller Control of the Housing Process*, New York: Macmillan, pp. 122–47.

—— (1976) *Housing by People*, London: Marion Boyars.

—— (1986) "Future Directions in Housing Policies," *Habitat International* 10(3): 7–25.

Underwood, G.C. (1986) "Zimbabwe's Urban Low Cost Housing Areas: A Planner's Perspective," *African Urban Quarterly* 2(1): 24–36.

UNDP (United Nations Development Programme) (1996) *Human Development Report 1996*, New York: Oxford University Press.

UNICEF (United Nations Children's Fund) (1987) *Adjustment with a Human Face*, New York: United Nations.

United Nations (1951) *Measures for the Economic Development of Under-Developed Countries*, New York: United Nations.

—— (1986) *Urban and Rural Population Projections, 1950–2025*, New York: United Nations.

—— (1996) *An Urbanizing World: Global Report on Human Settlements*, New York: Oxford University Press (published for the United Nations)

—— (1997) *Adjustment with a Human Face*, New York: United Nations Press.

—— (2003) *The Challenge of Slums: Global Report on Human Settlements 2003*, London: Earthscan Publications.

—— (2005a) *Report of the Fact-Find Mission to Zimbabwe to Access the Scope and Impact of Operation Murambatsvina by the United Nations Special Envoy on Human Settlement Issues in Zimbabwe*, Nairobi: United Nations.

—— (2005b) *Human Development Report 2005*, New York: United Nations Development Programme.

USAID (United States Agency for International Development) (1984) *Zimbabwe: Country Development Strategy Statement*, Washington DC: USAID.

—— (1985) *Housing Finance in Zimbabwe*, Washington DC: USAID.

Viner, J. (1953) *International Trade and Economic Development*, Oxford: Oxford University Press.

de Vries, B.A. (1996) "World Bank Focus on Poverty," in J.M. Griesgraber and B.G. Gunter (eds) *The World Bank: Lending on a Global Scale*, London: Pluto Press.

Wade, R. (2001) "Showdown at the World Bank," *New Left Review* 7: 124–137.

Wakely, P. (1999) "Urban Housing: The Need for Public Sector Intervention and International Cooperation," *Cities*, August: 195–201.

Wallerstein, I. (1974) "The Rise and Future Demise of the World Capitalist System: Concepts for Comparative Analysis," *Comparative Studies in Society and History* 16: 387–415.

—— (1976) *The Modern World System: Capitalist Agriculture and the Origins of the European World Economy in the Sixteenth Century*, New York: Academic Press.

—— (1979) *The Capitalist World-Economy*, New York: Cambridge University Press.

—— (1980) *The Modern World-System II: Mercantilism and the Consolidation of the European World-Economy, 1600–1750*, New York: Academic Press.

Walton, J. (1987) "Urban Protest and the Global Political Economy: The IMF Riots," in M.P. Smith and J.R Feagin (eds) *The Capitalist City: Global Restructuring and Community Politics*, London: Basil Blackwell, pp. 364–86.

Ward, B. (1962) *The Rich Nations and Poor Nations*, New York: W.W. Norton.

—— (1965) *The Decade of Development – A Study in Frustration?*, London: Overseas Development Institute.

Watts, M. (1983) *Silent Violence: Food, Famine and Peasantry in Northern Nigeria*, Berkeley: University of California Press.

Weaver, J.H. (1965) *The International Development Association: A New Approach to Foreign Aid*, New York: Praeger Publishers.

Weiss, R. (1994) *Zimbabwe and the New Elite*, London: British Academic Press.

Weissman, E. (1978) "Human Settlements – Struggle for Identity," *Habitat International* 3(3/4): 227–41.

Wekwete, K.H. (1988) "The Development of Urban Planning in Zimbabwe," *Cities*, February: 57–71.

Wellings, P. (1982) "Aid to the South African Periphery," *Applied Geography* 2: 267–90.

Wilber, C.K. (ed.) (1979) *The Political Economy of Development*, New York: Random House.

Williams, D.G. (1984) "The Role of International Agencies: the World Bank," in Geoffrey K. Payne (ed.) *Low-Income Housing in the Developing World: The Role of Sites and Services and Settlement Upgrading*, Chichester: John Wiley and Sons, pp. 173–85.

Williamson, J. (1990) "What Washington Means by Policy Reform," in J. Williamson (ed.) *Latin American Adjustment: How Much Has Happened?*, Washington DC: Institute for International Economics.

—— (2004) A Short History of the Washington Consensus, paper commissioned by Fundacion CIDOB for a conference "From the Washington Consensus towards a new Global Governance," Barcelona, September 24–25, 2004.

Wolfensohn, J. (1999) *The Comprehensive Development Framework, Draft*. Washington DC: World Bank.

World Bank (1950–1) *Sixth Annual Report*, Washington DC: World Bank.

—— (1965) *India's Economic Development Effort, Report to the President*, Washington DC: World Bank.

—— (1972) *Urbanization (Sector Paper)*, Washington DC: World Bank.

—— (1974) *Sites and Services Projects*, Washington DC: World Bank.

—— (1975a) *The Task Ahead for the Cities of the Developing Countries*, World Bank Staff Working Paper No. 209, Washington DC: World Bank.

—— (1975b) *Housing: Sector Policy Paper*, Washington DC: World Bank.

—— (1976) *Urban Poverty Action Program: Interim Report of the Urban Poverty Task Force*, Washington DC: World Bank.

—— (1981) *Accelerated Development in Sub-Saharan Africa: An Agenda for Action*, Washington DC: World Bank.

—— (1984a) *Principles of Adjustment Lending*, Washington DC: World Bank.

—— (1984b) *Staff Appraisal Report, Urban Development Project Zimbabwe*, Washington DC: World Bank (Eastern Africa Projects Department).

—— (1985a) "Making Shelter Projects Replicable," *Urban Edge* 9(10), 1–3.

—— (1985b) *Zimbabwe: Urban Sector Review*, Washington DC: World Bank (Eastern Africa Projects Department).

—— (1987) "Zimbabwe: A Strategy for Sustained Growth," Washington DC: World Bank (Southern Africa Department, Africa Region).

—— (1989a) *FY89 Sector Review, Urban Operations Department: Reaching the Poor Through Urban Operations*, Report INU-OR 3. Washington DC: World Bank (General Operations Review, Urban Development Department).

—— (1989b) *From Crisis to Sustainable Development*, Washington DC: World Bank.

—— (1989c) *Developing the Private Sector: Challenges for the World Bank Group*, Washington DC: World Bank.

—— (1989d) *Adjustment Lending, How It has Worked, How it Can be Improved*, Washington DC: World Bank.

—— (1990) *World Development Report (Poverty)*, Washington DC: World Bank.

—— (1991a) *Urban Policy and Economic Development: An Agenda for the 1990s*, Washington, DC: World Bank.

—— (1991b) "World Bank Shifts Urban Policy," *Urban Edge* 15(2), 1, 2, 6, 7.

—— (1991c) *World Bank (IBRD) and IDA Operations (Zimbabwe)*, Washington DC: World Bank (Eastern Africa Projects Department).

—— (1992) *The Urban Development Division* (pamphlet), Washington DC: World Bank (Urban Division).

—— (1993a) *Housing: Enabling Markets to Work*, Washington, DC: World Bank.

—— (1993b) *Dynamics of Changing Cities (Aide Memoire)*, Washington DC: World Bank (South Africa, Urban Economic Mission).

—— (1994a) *Twenty Years of Lending for Urban Development 1972–1992*, Report 13117. Washington DC: World Bank (Operations Evaluations Department Review).

—— (1994b) *Implementation and Completion Report, Urban Development Project*, Washington DC: World Bank (Eastern Africa Projects Department).

—— (1995) *Project Completion Report: Zimbabwe, Infrastructure Operations Division*, Washington DC: World Bank.

—— (1996) *World Bank Annual Report*, Washington DC: World Bank.

—— (2000a) *Cities in Transition*, Washington, DC: World Bank.

—— (2000b) *Implementation and Completion Report, Urban Sector and Regional Development (Urban II) Project*, Washington DC: World Bank (Eastern Africa Projects Department).

—— (2000c) *World Bank Annual Report*, Washington DC: World Bank.

—— (2000d) *Executive Briefing: Global Capital Markets Development*, Washington DC: World Bank.

Zinyama, L.M., Tevera, D.S., and Cumming, D. (1993) *Harare: The Growth and Problems of the City*, Harare: University of Zimbabwe Publications.

Zvobgo, E.J.M. (1981) "Foreword," in D. Patel and R.J. Adams (eds) *Chirambahuyo: A Case Study in Low Income Housing*, Gweru: Mambo Press, pp. ix–x.

Index

9 781138 987357